THE CILIATES

Biological Sciences

Editor

PROFESSOR A. J. CAIN
M.A., D.PHIL.
Professor of Zoology
in the University of Liverpool

THE CILIATES

A. R. Jones
Lecturer in Zoology
University of Reading

HUTCHINSON UNIVERSITY LIBRARY
LONDON

HUTCHINSON & CO (*Publishers*) LTD
3 Fitzroy Square, London W1

London Melbourne Sydney Auckland
Wellington Johannesburg Cape Town
and agencies throughout the world

First published 1974

*This book has been set in Times type, printed in Great Britain
on smooth wove paper by The Camelot Press Ltd, London and Southampton,
and bound by Wm. Brendon, Tiptree, Essex*

ISBN 0 09 117300 0 (cased)
0 09 117301 9 (paper)

DEDICATED TO MY
PARENTS

CONTENTS

PLATES

PREFACE

Although in the last decade there has appeared a number of textbooks on protozoology as well as many excellent monographs and reviews on particular aspects of these animals, there is no elementary text on the ciliate Protozoa. It is hoped that this little book will fill the gap. However, even a simple book of this sort cannot cover the whole field and the more informed reader will discover omissions both in depth and breadth. To alleviate this problem I have referred, as often as possible, to review articles or monographs. Sometimes I have done this in preference to quoting the original source.

Many people have helped in the writing of this book. Dr M. A. Sleigh and Professor J. A. Kitching, F.R.S., read and commented on a number of chapters; Dr L. H. Bannister provided both illustrative material (Plate 1) and illuminating conversation, and Dr J. Grainger introduced me to the secrets of activated sludge. Any errors which remain must of course be my responsibility. For an all too short time Miss R. Ash helped with the line drawings; she is responsible for Figs. 8, 9, 10, 12, 13, 21, 22, 28, 30, 37. Miss M. G. Smith bore the brunt of the typing ably assisted by Mrs P. M. Brown and Mrs G. I. Smillie. My wife gave both technical and moral support throughout and but for her this book may never have been started let alone finished. I should also like to thank many of my students whose freshness of outlook has helped me to see the wood for the trees. As one of them prefaced her report on food vacuole formation in *Spirostomum*: 'God hath chosen the foolish things of this world to confound the wise; and God hath chosen the weak things of the world to confound the things which are mighty'. No doubt ciliates will continue to confound and fascinate us for many years to come.

Reading   A. R. JONES

ACKNOWLEDGEMENTS

I am grateful to the following journals and publishers for the use of copyright material: Academic Press for Fig. 35D; *Biological Bulletin* for Fig. 27; *Journal of Cell Biology* for Fig. 10; *Journal of Cell Science* for Figs. 18, 19; *Journal of Experimental Biology* for Figs. 20, 24, 25, 26; *Journal of the Faculty of Science of the University of Tokyo* for Fig. 17; *Journal de Microscopie* for Fig. 9; *Journal of Protozoology* for Plate 3 and Figs. 11, 21, 22, 28; Oxford University Press for Figs. 1D, 2B and C, 3A and B, 4B, 5B and D, 31; Pergamon Press for Figs. 2D, 5A, B and C, 6, 7, 23, 30, 35A, B and C, 36, 37; *Philosophical Transactions of the Royal Society* for Fig. 13; *Symposia of the Society for Experimental Biology* for Figs. 14, 16; John Wiley & Sons Inc. for Fig. 34.

GLOSSARY

This glossary attempts to give brief definitions of words which either have a special meaning in, or are peculiar to, ciliate biology. It has drawn extensively on Corliss' (1959) 'definition of terms' included with his key to the higher groups of ciliates.

ADORAL ZONE OF MEMBRANELLES (AZM) An orderly arranged group or line of membranelles associated with the oral area of some holotrich and all spirotrichs, typically located on the left side of the buccal cavity.

ALVEOLUS See PELLICLE.

ANLAGE (plural—AGEN) An early stage in the development of an organelle or organellar complex. Frequently applied to the developing macro-nucleus and oral structures.

AUTOGAMY A pseudosexual process in which the karyological events leading to conjugation are performed but the gametic nuclei, rather than being exchanged with one from a partner, come together and fuse soon after their formation (see p. 135).

AXENIC CULTURE A culture in which there are no other organisms than the one being cultured; most importantly free of bacteria.

CILIA The whip-like organelles with which ciliates propel themselves and create feeding currents. Typically ciliates have *oral* (or *buccal*) *cilia* around the mouth, which are specialized to a greater or lesser degree for feeding, and *somatic cilia* over the general body surface for locomotion.

CIRRUS A structure composed of a number of somatic cilia acting together; oval or circular in cross section.

CONJUGATION A form of sexual reproduction unique to ciliates. Two gametic nuclei are produced by each of the two conjugants. The conjugants then exchange one of their nuclei (the migratory nucleus) which fuses with the stationary nucleus (see p. 131). In some cases the conjugants are specialized for their function and may differ markedly in size. The larger macroconjugant is usually sessile while the small microconjugant is free-swimming (see p. 134).

CONTRACTILE VACUOLE   Vacuoles in the ciliates' cytoplasm which gradually grow (diastole) and then contract (systole) emptying their contents to the exterior through a contractile vacuole pore.

CORTEX   The outer portion of the animal's cytoplasm containing the pellicle, cilia, infraciliature and a variety of other structures.

CYTOPHARYNX   The non-ciliated passageway leading from the cyto-stome into the inner cytoplasm of the organism.

CYTOSTOME   The cell mouth. The region in which food vacuoles are formed and material passes from being outside the animal to being inside it.

DEFINED MEDIUM   A culture medium containing only known chemicals. This may be a *minimal* medium containing only those chemicals essential for the growth of the organism being cultured.

DESMODEXY, RULE OF   This rule states that kinetodesmata always lie to the animals' right of the kinetosomes.

EDTA   Ethylenediamine tetra-acetic acid, a substance that chelates a variety of divalent ions including calcium.

EPIPLASM   See PELLICLE.

FOOD VACUOLE   An intracellular vacuole formed at the cytostome and containing food material. Digestion occurs in this vacuole when enzymes are added to it from primary lysosomes. The vacuole discharges its undigested contents through a cytopract.

GULLET   An ill-defined term frequently applied to the vestibule-buccal cavity region of the oral apparatus. Often used in descriptions of *Stentor* (see Fig. 37).

GLYCERINATED MODEL   See footnote, p. 78.

INFRACILIATURE   The subpellicularly located system intimately associated with the external ciliature both somatic and oral. It comprises mainly the kinetosomes and their associated fibre systems.

KINETODESMA (plural—MATA)   Longitudinally oriented fibrils associated with the kinetosomes. They lie to the animal's right of the kinetosomes and run anteriorly.

KINETOSOME (BASAL BODY)   That part of the cilium lying below the level of body surface It can be visualized in the light microscope by silver staining. This portion of the cilium persists even when the shaft has been lost or shed.

KINETY   A given row of kinetosomes with its associated kinetodesmata.

LORICA   A test or shell, often vase shaped, secreted by the animal. Largely organic, it may have foreign matter included in it.

MACRONUCLEUS   A polyploid (rarely diploid) nucleus largely responsible for all nuclear functions except those involved in sexual reproduction. (See also MICRONUCLEUS.)

MATING TYPES   Subspecific groups within which individuals are non-compatible for conjugation. These mating types are grouped into syngens or varieties. Conjugation only takes place between animals of different mating types but the same variety. At present several

varieties may be given the same specific name but there is good cause
to regard varieties as species (see p. 135).

MEMBRANELLE   A structure composed of a number of oral cilia acting
as a unit; usually rectangular in cross section and organized into an
adoral zone of membranelles (AZM).

METACHRONY   Coordinated movement by cilia in which they all beat
with the same frequency and in phase with their neighbours to either
side but slightly out of phase with those before and behind. For a
more detailed description and variations see p. 69.

MICRONUCLEUS   Diploid nucleus of ciliates which seems to play little
or no part in the life of the organism except at sexual reproduction,
when it undergoes meiosis to give rise to gametic nuclei.

MICROTUBULES   Proteinaceous tubules about 24 nm in diameter. They
are important elements in many ciliate organelles and structures
including cilia and cortical fibres.

MONILIFORM   Having a shape suggesting a string of beads.

MUCOCYST   A small pellicular organelle which apparently secretes mucus.

MYONEME   Contractile organelle functioning as a subcellular 'muscle'.
Those of *Stentor* are sometimes known as M bands.

NUCLEUS   See MACRONUCLEUS and MICRONUCLEUS.

OPISTHE   See PROTER.

PELLICLE   The outer covering of ciliates which has a typical structure
wherein the cell membrane is underlain by membrane-bounded
spaces called alveoli. The inner membranes of the alveoli are often
elaborated with fibrous material to form the epiplasm (see Fig. 9).

PINOCYTOSIS   The uptake of fluid through the formation of small
vesicles. This process is usually stimulated by substances in the
medium such as salts or aminoacids.

PRIMORDIUM   The embryonic stage in the development of organelles
or cortical structures. Frequently used in connection with the developing oral apparatus.

PROTER   When ciliates divide the anterior daughter cell is termed the
proter and the posterior one, the *opisthe*.

ROSETTE   The specialized structure appearing near the cytostome of
many apostomes.

SCOPULA   A region of specialized basal bodies that produce the stalk of
many ciliates including peritrichs and suctoria.

SYMBIOSIS   An association between two species of organism beneficial
to both. Such relationships occur between ciliates and algal cells,
the latter (endosymbionts) living in the cytoplasm. Ciliates also occur
in the gut of many herbivorous mammals where they may be involved
in a symbiotic relationship.

TELOTROCH   The free-swimming larval phase of sessilinid peritrichs.
As well as the oral cilia it has a ring of locomotory cilia, the trochal
band.

TRICHITES (RODS)   Skeletal elements in the oral apparatus of gymnostome (and rarely, hymenostome) ciliates, composed largely of
microtubules.

TRICHOCYST An extrusile organelle which discharges a thread. The thread may be armed with a pointed head or may contain toxic substances.

UNDULATING MEMBRANE (UM) A single line of cilia lying on the right border of the buccal cavity which together with the adoral zone of membranelles forms the oral ciliature.

UNITS μm, micrometre $= 10^{-6}$ metre (sometimes called a micron and designed μ); nm, nanometre $= 10^{-9}$ metre (1 nm $= 10$ Å).

VARIETY See MATING TYPES.

ZOOCHLORELLA The green algal cells that live in symbiotic relationship with some ciliates (and other animals).

ZOOID The 'body', as opposed to the stalk, of sessile ciliates, e.g. many peritrichs and suctoria.

INTRODUCTION

The last twenty years have seen a great upsurge of interest in Protozoa, especially ciliates. The main reasons for this revitalization are probably that the electron microscope, defined culture techniques and the availability of sophisticated biochemical apparatus have allowed us to look at structure and function with far greater definition than ever before. The current general interest in 'cells' has resulted in many workers being attracted by Protozoa, systems that so manifestly contain both cell and organism within a single membrane. This added knowledge of the biochemistry, physiology and ultrastructure has affected all who work with unicells and has permitted, or perhaps demanded, a rethink of the systematics of the group; ciliate taxonomy today is a healthy, lively and active science. Luckily during these two decades of furious expansion there have been ciliate biologists of sufficient intellectual strength to guide and direct the flood; without them chaos might easily have descended. Perhaps it is true to say that ciliate biology was preadapted for such expansion. Certainly excellent groundwork had been laid during the '30s and '40s by such people as Chatton and Lowff in France, Hall and Sonneborn in America, Kitching in England, Kinosita in Japan and many others. Perhaps the one to whom we most owe a debt of gratitude is Fauré-Fremiet, who died in 1971, after over 65 years of influential work in many fields of ciliate biology (Obituary, *Nature, London,* **236,** 43). Of the younger generation I am especially indebted to Pitelka, Corliss, Tartar, Zeuthen, Rudzinska and Sleigh to mention a few. No doubt each protozoologist has his own private list.

Although no general text devoted to ciliates has been published recently, information on the group is widely available. The standard textbooks of protozoology (Hall, 1953; Kudo, 1954) contain sections on ciliates. More recently Mackinnon and Hawes (1961) have dealt with a limited number of ciliate types in some detail. The English translation of Dogiel (1965) contains a considerable amount of information on ciliates and gives special emphasis to Russian work, often neglected in the West. Sandon's (1963) little book contains a brilliant essay on ciliates. Books dealing with particular topics of interest to the ciliate biologist include those of Corliss (1961: taxonomy), Pitelka (1963: fine structure), Tartar (1961: morphogenesis in *Stentor*), Sleigh (1962: cilia and flagella) and Hall (1965: feeding and nutrition). Of very great value to the researcher or student requiring more information are the collections of reviews of Protozoa that have been recently published. The forerunner of these volumes is to be found in *Protozoa in Biological Research* (Eds. Calkins and Summers, 1941). Then followed three volumes of *Biochemistry and Physiology of Protozoa* (vol. I, 1951, Ed. Lwoff; vol. II, 1955, Eds. Hutner and Lwoff; vol. III, 1964, Ed. Hutner). More recently have come the three volumes of *Research in Protozoology* (Ed. Chen, 1967a,b, 1969) and the volume of *Chemical Zoology* devoted to Protozoa (Ed. Florkin, Scheer and Kidder, 1967) which cover most of the more active fields of ciliate research.

I

FORM

There are about 5700 species of ciliate (Corliss, 1961).* One tends to think of these protozoa as being uniformly small but in fact they vary in size from about $10 \, \mu m$ to about $3000 \, \mu m$ maximum dimension. The largest forms are usually flattened or elongated but nonetheless the volume range may be from about $550 \, \mu m^3$ to $350 \times 10^6 \, \mu m^3$. Thus the size ratio between the largest and the smallest ciliate is about the same as that between a blue whale and a rat. It is not surprising to find within this range considerable variety of morphology (see Figs. 1–5). Despite these differences, however, there are many features common to most members of the group, and the class Ciliophora is generally agreed to be phylogenetically homogeneous. No single characteristic can be considered as diagnostic but there are six features most of which are possessed by all ciliates and which, taken together, clearly separate the ciliates from other protozoa (Corliss, 1961), even such superficially similar ones as the opalinids.

(1) Possession of cilia; at least during some part of the life-cycle.
(2) Possession of an infraciliature located in the subpellicular region composed chiefly of the ciliary bases (kinetosomes) and their associated fibre systems. The infraciliature persists throughout the life-cycle.
(3) Possession of two sorts of nuclei. The smaller micronucleus is usually diploid and the animal may have one or many, although amicro-nucleate strains are known. These nuclei are involved in sexual reproduction. The larger macronucleus is usually polyploid (but see p. 171) and plays no part in sexual reproduction.

*A classification of the subphylum Ciliophora is to be found in Chapter 9 (p. 162).

transverse

(4) At binary fission the division plane typically cuts across the longitudinal rows of cilia (Fig. 29).

(5) Sexual reproduction is typically conjugation involving the reciprocal exchange of gametic nuclei.

(6) Possession of a functional mouth often with associated oral structures of various origins. In some groups the mouth has been lost (e.g. Astomatida, Suctorida).

BODY SHAPE

The simplest and possibly the commonest body shape is that of an ovoid (Plate 4 and Fig. 1) or pear-shaped solid and, less frequently, a sphere. Very many of the smaller free-swimming ciliates are one of these shapes. They are usually fairly unspecialized, with uniform ciliation and with the mouth parts either terminal or placed laterally up to one half of the way down the animal's length. Those with apical mouthparts are most often found in the most primitive order, the Gymnostomatida, and are often very regular ovoids or spheres. They are usually carnivores (eating prey as large or larger than themselves) which swim along, their mouths before them until they collide with suitable prey. When this occurs engulfment starts immediately and is often completed quickly. A common example of these voracious animals is *Coleps* (Fig. 1A), often found in pond-water. *Coleps* is noteworthy in that not only is it a typical rhab-dophorine gymnostome but it also exhibits calcified skeletal elements. Some of its relations have evolved a more or less long and mobile proboscis from the anterior of the body. The mouth is placed at the base of the proboscis in many genera (e.g *Dileptus*) (Fig. 1B) but may be near its tip (e.g. *Lacrymaria*) (Fig. 1C). As they swim, the proboscis sweeps the water ahead increasing the chance of collision with suitable prey. The proboscis is often armed with stinging organelles called trichocysts. Less voracious are the cyrtophorine gymnostomes, mostly vegetarian. *Nassula* (Fig. 2A), for example, feeds on filamentous algae and in this case the mouth is not apical but laterally placed on its ovoid body, collision not being an element in its feeding. With ciliates of this shape that feed either on small particles, detritus or vegetation the mouth is often laterally placed and there may be elaborate oral cilia. The lateral placement presumably gives more protection to the delicate feeding organelles. A typical ciliate of this sort is *Tetrahymena pyriformis* (Fig. 2B). Sometimes the mouth is further protected by being on the base of a depression. The regular outline may be thus interrupted on one side by such a concavity and this results in animals with a hooked appearance (e.g. *Loxodes*, Fig. 7) or even kidney-shaped as in *Colpoda* (Fig. 1D).

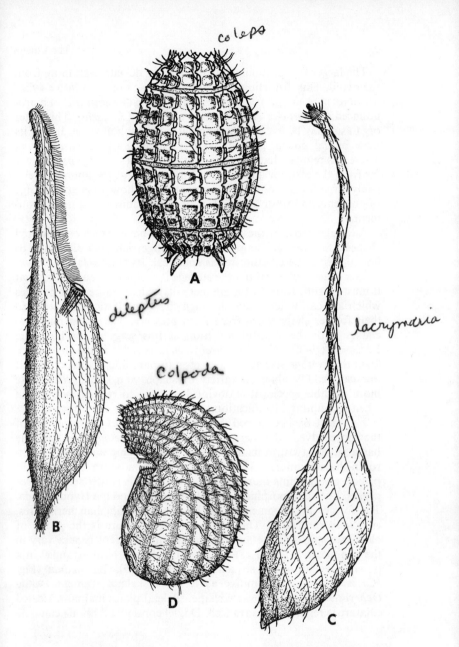

coleps

dileptus

Colpoda

lacrymaria

Fig. 1
A, *Coleps hirtus*, B, *Dileptus anser* and C, *Lacrymaria olor*, all <u>rhabdo-phorine gymnostomes</u>. D, the trichostome *Colpoda cucullus*. B, modified after Hall (1953); D, after Mackinnon and Hawes (1961).

ABC

The larger free-swimming ciliates tend to deviate even more from the ovoid. This distortion may simply be an elongation into a cylindrical form as in *Spirostomum* (Fig. 4A) or the development of a pronounced oral groove (*Paramecium*) or pouch (*Bursaria*). This latter organism may be over 1 mm long but is almost hollow because of its huge buccal pouch, large enough to engulf a whole *Paramecium*. These less regular shapes ensure that no part of the organism is very far from the exterior and that the surface area of the animal is relatively large for its volume. These properties allow a faster exchange of nutrients and waste materials between the animal and its environment.

Although most of the advanced ciliates have rather complicated shapes a few are very simple. Thus the oligotrich *Halteria* (Fig. 5B) has sophisticated ciliature but is spherical. Its close relation *Strombidium* is also spherical but the surface is sculptured, suggesting that it must be stiff. Most of the simpler ciliates have rather soft pellicles which are easily deformed although they are elastic and return to their former shape when freed from pressure. The shape may also change with the state of nutrition at this stage in the life-cycle. Furgason (1940) writes of *Tetrahymena,* 'so wide is the divergence from the average size and shape, even in pure line cultures . . . that one can readily obtain a variety of shapes and sizes'. With these more variable species it is obviously unwise to rely on body form when attempting identification.

The most bizarre-shaped free-swimming ciliates are members of the more advanced orders. The heterotrich *Metopus* for example has a violent twist in its anterior third producing an almost mushroom-shaped animal. The development of a more or less rigid pellicle also permits more irregular forms. This is best seen in three orders, Entodinomorphida, Odontostomatida and the Hypotrichida. The former are commensals in the gut of mammalian herbivores, especially ruminants. The most noticeable feature is the group of elaborate spines at the animal's posterior. Different genera vary in the complexity of their external form and have been arranged in a possible evolutionary sequence (Dogiel, 1965). *Entodinium* (Fig. 5C) is considered primitive as it lacks skeletal elements while *Ophryoscolex* with its ornate shape, skeletal plates and complicated ciliature is regarded as advanced. Dogiel considers that one can date

---

Fig. 2
A, *Nassula aurea*, a cyrtophorine gymnostome. B, *Tetrahymena pyriformis*, a tetrahymenid hymenostome. C, *Spirochona gemmipara*, a chonotrich. D, *Hemispeira asteriasi*, a thigmotrich. B, and C, modified after Mackinnon and Hawes (1961); D, after Corliss (1961).

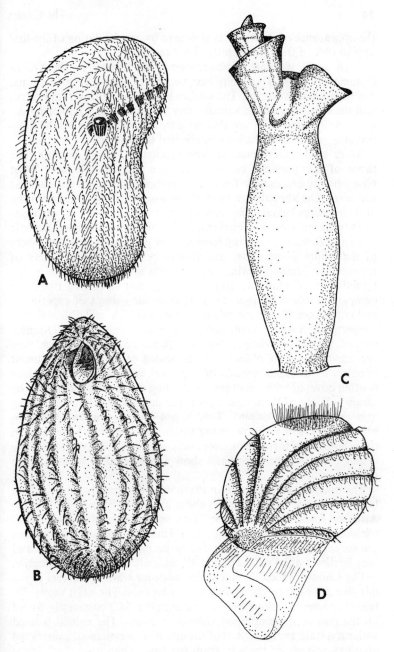

Fig. 2

the appearance of these different genera by careful dating of the first appearance of their hosts in the fossil record.

Even more extreme surface sculpturing occurs in the Odontostomes (Fig. 5A). These may have two lateral flaps giving the animal a carapace-like structure. The surface has various grooves and ridges and the two sides of the animals may be elaborated to differing degrees. Spines and teeth are also present. It is difficult to see the functional significance of such a complicated body form.

More easily interpreted are the hypotrichs. These animals, like those of the previous two orders, have reduced ciliature. What cilia remain are mostly fused into compound organelles such as stiff cirri. The animals are usually dorso-ventrally flattened with the oral structures below. They move over the surfaces of their environment 'walking' with their cirri in a well coordinated and characteristic scurrying fashion, gathering food as they go. They are the browsers of the ciliate world. They are able to perform a wide variety of movements: jumping, swimming forwards and backwards in a way reminiscent of some tiny crustacean. It is not surprising that their complex motile behaviour has long been the subject of experiment and conjecture. These animals have a firm pellicle and although this property has not been exploited to the same extent as in the Odontostomes, many have ridges, flaps and papillae. In dorsal view, they are mostly oval to oblong but elongated (*Uroleptus*) and almost circular forms (some species of *Euplotes*) exist. The dorsal surface is often covered with small pits containing single cilia thought to be sensory. Some of the longer hypotrichs are twisted about their long axis (e.g. *Hypotrichidium*). This is not an uncommon feature in other orders. For example, many heterotrichs show it, sometimes in the whole body form, sometimes only in a spiral disposition of the ciliary rows. All the apostomes show a great deal of torsion in the adult stage.

The most complicated and in many ways the most beautiful ciliate body forms are to be found in those animals which are sedentary. Attached forms are common in many suborders, and in the Chonotrichida, Suctorida, Peritrichida and Tintinnida most of the genera are sedentary. The attachment can be permanent or temporary and may be direct or by means of a stalk, which is sometimes contractile.

The Chonotrichida form a small suborder whose members typically live attached to the surface of crustaceans. The most easily obtained species is *Spirochona gemmipara* (Fig. 2C) commonly found on the gills of the amphipod *Gammarus pulex*. The animal is fixed without a stalk to the surface of the gill. It is vase-shaped, constricted about two-thirds of the way from the base. The upper third opens out into a complicated spirally-wound protoplasmic collar bearing cilia which create feeding currents. A marine genus, *Chilodochona,* lives on the maxillae and maxillipeds of crabs and has a much simpler collar, also a short non-contractile stalk. As with other sessile animals there must be a motile distribution phase in the life-cycle. This problem is dealt with on p. 145.

The suctorians are peculiar among ciliates in that the adults never have cilia or a mouth. They are included in the subphylum because of the obvious ciliate affinities of the larva, the persistent infraciliature and the mode of sexual reproduction. The animals are usually, but not always, borne upon a stalk which is non-contractile. A typical freshwater form is *Podophrya collini* (Fig. 3C) which can be found in ponds, sometimes hanging from the surface film. The body is spherical to pear-shaped with a rather obvious hyaline layer around it. From the body protrude the typical suctorian tentacles, perhaps as many as fifty of them, each with a small swelling at its tip making the whole animal look somewhat like a pin-cushion. In this species the tentacles are scattered over the surface but in many genera they are arranged into groups or restricted to one part of the body surface. The tentacles of most genera are knobbed but this is not always the case. Food is ingested through the tentacles (see p. 92) and the body swells; a heavily fed *Podophrya collini* may have a diameter more than three times that of a starved animal. In such sated animals the tentacles are usually shorter whereas in starved individuals they are extended. Suctorian body shape is very variable and although many genera follow the simple pattern of a round or conical zooid on a stalk, other genera may lack a stalk or have a zooid of complex shape. Some of the stalkless forms (*Dendrosoma*) are dentritic, looking almost like tiny corals while *Lernaeophrya* forms an encrusting growth on hydrozoan colonies. The hemispherical *Dendrocometes* adheres to the gills of *Gammarus,* and is not only stalkless but also has branched spiky tentacles. So bizarre are these animals that many almost defy description and it is suggested that the reader refers to the figures in any of the more comprehensive protozoology textbooks (Hall, 1953; Kudo, 1954; Wenyon, 1926). Many suctorians have a hard extracellular sheath called a lorica. This is not usually the same shape as the animal itself but often vase-shaped and loosely enclosing the organism so that only the tentacles protrude (*Thecacineta*). Such a protective structure is common in sessile ciliates and will be met with in other suborders. There can be few animals with such quiet beauty combined with such peculiar feeding habits and life-cycle. The aesthetic appeal of these creatures is best illustrated in the photographs, films and drawings of the German protozoologist Grell (1968).

The best known of the sessile ciliates are the Peritrichs. These

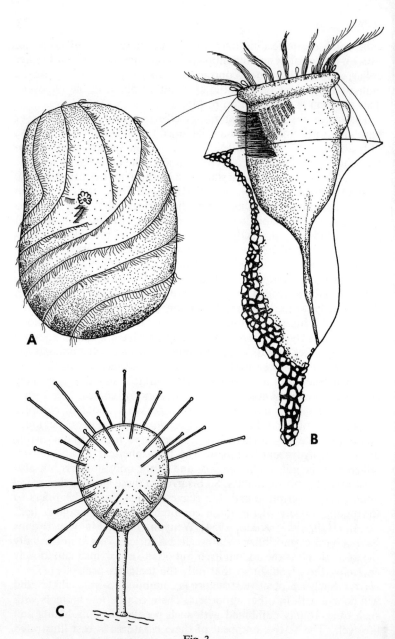

Fig. 3
A, *Gymnodiniodes inkystans*, an apostome. **B,** *Tintinnopsis campanula*, a tintinnid. **C,** *Podophrya collini*, a suctorian. **A** and **B,** after Mackinnon and Hawes (1961).

animals are common in both marine and freshwater habitats and were first observed in 1677 by Leeuwenhoek in an infusion of pepper (Dobell, 1932). Many protozoologists have been attracted to these animals as much by their aesthetic appeal as anything else, for as Kent (1880) has written: 'the almost constant state of alternate contraction and prolongation . . . in connection with a large colony of these elegant microscopic beings produces a spectacle of unparalleled beauty and activity'. *Vorticella* (Fig. 4C) has long been considered the typical genus. A typical species however is not to be found as the genus contains over 200, many of which differ only in minutiae. Noland and Finley (1931) have provided a review of the genus and a detailed description of a few of its species. The account which follows is a general one. The animal is divided into a stalk and a zooid. It is solitary. The stalk is hyaline with a rather more opaque thread running up inside it in a loose spiral. This thread, the myoneme, can be seen to be enclosed in a canal within the stalk. The myoneme is a muscle-like organelle, responsible for the rapid contraction of the stalk. The zooid is roughly conical attached at its apex to the stalk. This, too, has myonemes which are more difficult to see than those of the stalk. Contraction of these causes a shortening of the cone and a tucking in of the oral structures and results in the zooid becoming almost spherical. When relaxed, however, the base of the cone can be seen to bear the complicated wreaths of oral ciliature, the only cilia that the adult possesses. The pellicle of the zooid often has circumferential striations or small papillae.

Many genera of peritrichs are not solitary but colonial. One such creature is *Carchesium polypinum*, a common inhabitant of ponds. After the larva has settled it produces a long (perhaps 2 mm) stalk and then divides many times to produce an arborescent colony. The stalks of such a colony are contractile. In the case of *Carchesium* the myoneme is not continuous through all the branches but is interrupted at each bifurcation and each zooid and its stalk is capable of independent contraction. In *Zoothamnium,* however, the myoneme is continuous throughout the colony and contraction involves the whole colony. Other genera, both solitary and colonial, have stalks which are not contractile although the zooids have myonemes. *Campanula* and *Opercularia* are two common examples.

Asexual reproduction of stalked peritrichs is by the production of a motile larva, the telotroch (see Chapter 7). The suborder Mobilina contains genera that appear to have arisen by neoteny from the stalked form. That is to say, the larval condition persists and reproduction occurs during the larval life, thus producing animals without a sessile stage. These organisms look much like telotroch larvae and the group includes *Trichodina* (Fig. 4B), some species of which are

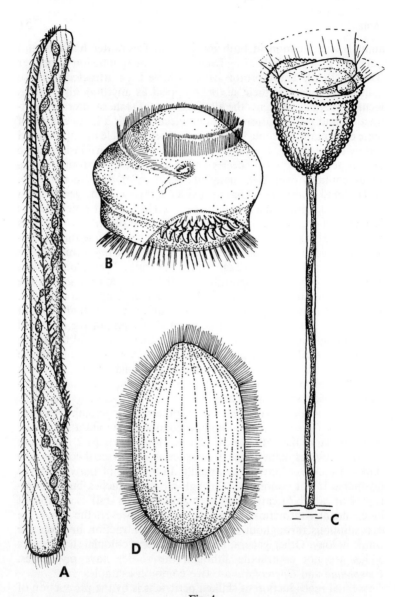

Fig. 4
A, *Spirostomum ambiguum*, a heterotrich. B, *Trichodina myicola*, a mobilinid peritrich. C, *Vorticella monilata*, a sessilinid peritrich. D, *Anoplophrya marylandensis*, an astome. B, after Mackinnon and Hawes (1961); D, modified after Kudo (1954).

important ectoparasites of fish. The aboral surface is flattened to form a disc which is equipped with a series of 26–36 denticles arranged radially and articulating with one another. It is thought that these play some part in the attachment to the host.

Some sessile peritrichs have a lorica into which they can contract (e.g. *Vaginicola*). In these forms the stalk is often lacking or reduced although the lorica itself may be borne on a stalk. Some species are equipped with a horny disc which acts as an operculum to plug the mouth of the lorica when the animal contracts, while others have flexible valves around the mouth of the lorica that close the opening when the occupant withdraws inside. In some cases the lorica may be fixed to the substrate over a wide area (e.g. *Lagenophrys*), while in others the point of attachment may be small. An interesting case of convergent evolution is found between these loricate peritrichs and a group of heterotrichs, the folliculinids. These latter are also loricate, the lorica being attached for some part of its length. The animal is contractile and can retract inside the lorica and its oral ciliature is developed on the large bilobed anterior end to produce a feeding current. Another heterotrich group, the lichnophorids, are not unlike the stalked peritrichs in functional morphology although they are not at all closely related. *Lichnophora,* for example, has a ring of oral cilia and a curiously well developed foot by which it attaches itself to algae and a variety of marine animals, including starfish. Yet another heterotrich, the well-known *Stentor* (Fig. 30), has functional similarities to the vorticellids. *Stentor* spends much of its time fixed to the substrate but is able to detach and swim away. The whole animal is strongly contractile. When attached, as the name suggests, it has the shape of a coaching horn. The well developed oral ciliature is arranged in a ring around the wide anterior end. When swimming freely the shape is more that of a blunt cone and from either of these conditions the animal is able to contract to a sphere. There is a contractile system that causes the oral ciliature to be covered and protected. Thus both *Stentor* and *Vorticella* feed while extended on a long stalk or stalk-like region, pulling food towards themselves by means of a circularly dispersed set of vortex-creating cilia borne on a wider anterior end or zooid. Both animals can bend from side to side or contract down to their point of attachment. It is interesting to note that in some species of *Stentor* a mucus 'protolorica' is formed into which the animal can contract.

The members of the order Tintinnida are also sessile and loricate. The lorica in this group is sticky and small particles of sand and detritus adhere to it. Some members of this group are pelagic or attached to the water surface. The body shape, as with most attached ciliates, is that of a cone with elaborate oral cilia at the broader end.

CILIATURE AND ORAL APPARATUS

Our knowledge of the arrangement of cilia has been much advanced by the technique of silver staining. First Klein, then Chatton and Lwoff (1935) and more recently Corliss (1953) have adapted and improved the methods. The silver depositions occur on or near the kinetosomes (see Chapter 2). Silver is also deposited on the so-called silver-line system, a pellicular network (Plate 4). These markings and, more especially, the arrangement of the ciliary bases are of great taxonomic importance.

The ciliature can be divided into two types. Firstly, there are the somatic cilia usually arranged in longitudinal rows or kineties. In unspecialized animals these cilia are distributed more or less evenly over the body surface and their main function is locomotion. Secondly, there are the oral cilia. These are associated primarily with feeding and can in turn be divided into two separate sorts, homologies of which can be traced in most ciliates. The ciliature of *Tetrahymena pyriformis* (Plate 4 and Fig. 2B) provides a useful general example of the ciliary organization. As its name implies, this animal is pear-shaped. The somatic cilia are distributed over the surface in 15–24 kineties. The number of kineties varies with the strain but even within a strain it is not very constant. Thus *Tetrahymena pyriformis* strain GL has 15–20 (mean 17) kineties (Loefer *et al.*, 1966). They run longitudinally towards the poles. Two kineties, designated 1 and 2, are interrupted by the oral apparatus and terminate there. At the anterior pole the kineties either terminate at a circumpolar silver-staining ring, the apical loop, or converge on a line running from the oral structures to the loop. At the posterior end the kineties stop short of the pole and thin silver-staining lines continue on to convergence. There are about 70 cilia in each kinety giving a total of about 1200.

The oral ciliature is localized on the edge of, or in, a depression in the body surface known as the buccal cavity. This cavity is also pear-shaped and aligned along the animal's long axis. It has a pronounced ridge along its left side. Running along its right side is a single line of cilia set closely together. The shafts of the cilia appear to be loosely attached to each other. This structure is in almost constant motion and is known as the undulating membrane (UM) and constitutes one part of the oral ciliature. The second part is made up of the membranelles. These structures, of which *Tetrahymena* possess three, are groups of cilia acting as a single unit. The cilia that make up these organelles usually arise from 3 or 4 closely set rows of kinetosomes which give the structure its rectangular ground-plan. The three membranelles of this species make up the adoral zone of membranelles (AZM). As will be seen, many ciliates have a more extensive AZM. The most anterior of *Tetrahymena*'s membranelles is the longest and is situated on the upper left edge of the buccal cavity beating downwards and inwards. The second and third are placed further in the buccal cavity and are smaller. At the bottom of the buccal cavity food particles are accumulated and food vacuoles are formed that pass into the organism at the cytostome.

The arrangement of cilia in *Tetrahymena* although simple is not the most primitive. For that we must look to rhabdophorine gymnostomes such as *Holophrya* and *Prorodon*. Both are nearly spherical with meridional kineties. The mouth is at the anterior pole. There is no buccal cavity, the cytostome being on the surface of the organism and is supported by intracellular rods or trichites characteristic of the order. There is no oral ciliature. In the cyrtophorine gymnostomes the somatic ciliature may be restricted to certain parts of the animal and the oral apparatus is complex (e.g. *Nassula*) but there is no oral ciliature.

In the other lower ciliate orders, the Trichostomatida and the Chonotrichida, there are some quite specialized forms. The trichostome *Colpoda* is common in freshwater and soil. The kineties are set in grooves and the cilia occur not regularly spaced out but in pairs, for most of the length of the kinety. The cytostome is set in a depression in the body surface, the vestibulum, which bears cilia somewhat specialized for feeding. There is however no sign of either UM or AZM. The chonotrichs are largely devoid of cilia except on the complicated collar which surrounds the large vestibulum. On this a few rows of somatic cilia are found. The motile larva is ciliated and bears some resemblance to the cyrtophorine gymnostomes in its ciliation (and morphogenesis).

Also lacking oral ciliation are the members of the three orders Apostomatida, Astomatida and Suctorida. The first of these groups contains animals that are parasites or commensals of a variety of marine organisms, mainly crustaceans. The oral apparatus consists of a small cytostome with the enigmatic, possibly sceretory, rosette nearby. The somatic ciliature of the adult is arranged in kineties that spiral round the animal, sometimes in a rather complicated way. This ciliary pattern is simplified prior to division, which usually takes place in a cyst. The larval young are distributive. Six morphologically distinct stages are recognized in the life-cycle. Chatton and Lwoff (1935) reviewed the complicated reproductive biology of these animals and discussed the theoretical problems that it raised.

The astomes have no oral ciliature for the simple reason that they have no mouths. They are mainly parasites and commensals in a variety of invertebrates, especially oligochaetes. There seems little doubt that this group is polyphyletic and contains degenerate members of other orders adapted to their parasitic mode of life. De Puytorac (1959) is of the opinion that they are mostly thigmotrichs with possibly a few apostomes and tetrahymenids. The somatic ciliature is usually uniform over the body surface in meridional or slightly helically disposed kineties.

The suctorians are perhaps the most extraordinary as ciliature, both somatic and oral, is lacking in the adult. However, silver staining shows that kinetosomes persist throughout the life-cycle. The ciliated larva lacks a mouth and it is difficult to establish affinities with other orders. In some cases, e.g. *Ephelota* (Guilcher, 1948), there may be a vestigial oral aperture but there is no sign of oral ciliature. The somatic ciliature of the larva is more or less uniform but may be confined to its lower surface. The cilia are arranged in rows which sometimes arch around the anterior scopula, a circular area of close-set cilia which is the site of the stalk production at metamorphosis.

True oral ciliature makes its first appearance in the hymenostomes. The typical organization for this order has already been described. It is interesting to note that the tetrahymenid *Pseudomicrothorax* possesses both the UM–AZM typical of its order and also cyrtophorine-like trichites suggesting a strong evolutionary link between the gymnostomes and the tetrahymenids. *Paramecium* was long considered not to have specialized oral ciliature, but this is now thought incorrect. The oral apparatus is at the end of a groove which extends from the front of the animal. This groove leads to a wide shallow vestibulum, containing somatic ciliature, which narrows to form a buccal cavity in which are the oral cilia. These are difficult to interpret but Coliss (1956) equates the endoral membrane with the UM. The dorsal and ventral periculi and the quadrulus are thought to be three membranelles thus giving *Paramecium* a typical set of tetrahymenid oral cilia. Other members of the order, e.g. *Pleuronema,* have a very well developed UM while the membranelles are relatively small. The buccal cavity is large but shallow and somatic cilia are mainly responsible for passing particles of food to the UM which is held out like a net. The particles are trapped on the UM and pass to the cytostome. Much of the somatic ciliature of this animal is sparse and it holds itself to the substrate by the thigmotactic and adhesive dorsal cilia while it feeds. There are also stiff cilia which protrude bristle-like from the posterior of the animal. When it swims the UM is retracted, to be flicked out again when the animal settles.

The peritrichs have a highly specialized ciliature. The somatic ciliature is suppressed except in the larval stage and the neotenous

forms where a posterior ring of locomotory cilia is strongly developed. The oral cilia are well developed but the homologies with the
AZM–UM are difficult to determine. The cilia are arranged in a
band around the edge of the circular peristome, running round a
full turn before plunging down into the infundibulum at the base of
which is the cytostome (Fig. 13C). This band of cilia is made up of 1
or 2 inner rows and a single outer one. Although the bases of these
cilia may be fused the shafts are not and there are no membranelles.
The inner rows stand more or less upright and cause food particles
to be carried on to the horizontal outer row whence they are swept
into the infundibulum.

In the subclass Spirotricha, the oral ciliature, especially the AZM,
is much developed. The somatic ciliature, though sometimes complete and uniform, is often reduced and specialized. *Spirostomum
ambiguum* (Fig. 4A) is a relatively simple heterotrich. The somatic
ciliature is arranged in kineties which loosely spiral around the
animal. The AZM is composed of many membranelles and runs
from the anterior of the animal down two-thirds of its length to
terminate at the buccal cavity. The UM appears to be absent. In
Stentor also the UM is missing. In this genus the AZM describes a
part circle around the flat open peristome region, then twists down
into the closed funnel-like part of the buccal cavity. These membranelles beat so as to produce a vortical current which sweeps
particles down into the buccal cavity. The somatic ciliation of
Stentor also has a complex arrangement. The animal is cone-shaped
and there are fewer kineties at the narrow posterior end than at the
anterior end. This is due to the branching of kineties anteriorly and
the cessation of some kineties before they reach the posterior end.
The distance between the kineties increases in an orderly way round
the body of the animal. The meridians below the buccal cavity are
closest together and the space between them increases from left to
right so that the widest intermeridional space comes to lie next to the
narrowest (Fig. 30). This area is known as the locus of stripe-width
contrast (Tartar, 1961). As the animal grows, new kineties are added
in this area which also plays an important part in the initiation of the
oral cilia of the daughter cell. There are also somatic cilia in the
peristome region which derive from lateral body wall at division. The
kineties here are arranged in almost full circles running parallel to
the AZM.

Although absent in *Stentor* and *Spirostomum* the UM is to be
found in some heterotrichs, e.g. *Nyctotherus,* a parasite of frogs.
Even in this case, however, it is reduced and nothing like as powerful
as the AZM.

The dominance of the AZM is also seen in the oligotrichs. In this order somatic ciliature may be entirely lacking, or reduced to a number of bristles. The membranelles, although few in number, are large and powerful and play a part in locomotion as well as in feeding. The UM, although much reduced, is present. The ciliature of the tintinnids is similarly organized. There is 12–24 strong membranelles arranged around the peristome, which produce vortical feeding currents. The cilia making up the inner side of the membranelles tend to 'fray' apart so that even with the light microscope it is possible to detect the compound nature of these organelles. The somatic ciliature is much reduced. In *Tintinnopsis* there is a small patch of cilia on the opposite side of the animal to the buccal cavity, one line of which is elongated into a membrane-like structure. These cilia have been implicated in lorica construction and cleaning (Fauré-Fremiet, 1924). There are also structures, said to be cilia, which are attached to the edge of the peristome and whose distal ends apparently adhere to the edge of the lorica (Fig. 3B). These so-called 'elastic' cilia are supposed to support the animal after the fashion of tent guy-ropes as it extends out of its lorica.

Hypotrichs also have a very well developed AZM (Fig. 5D), but in this group the somatic ciliature emerges with a new emphasis. Almost all the body cilia have here been combined into a relatively small number of compound organelles, cirri. These are mainly distributed on the ventral side of these flattened animals. *Euplotes eurystomus* for instance has a total of 18 cirri arranged in a definite pattern of 9 fronto-ventrals, 5 anals and 3 caudals (Gliddon, 1966). These cirri are capable of coordinated movements during walking and swimming. The latter activity is aided by the AZM of 40–50 membranelles which runs along the anterior margin of the animal before curving down the animal's left ventral surface to end in the buccal cavity. The AZM delineates 2 sides of the triangular and somewhat concave peristome. The dorsal side bears 7 rows of simple bristle cilia which arise from small depressions in the pellicle. A single row of bristle cilia is found on the left ventral surface. The UM is very much reduced.

Most of the entodiniomorphids have no somatic cilia or UM. There are often two clumps of membranelles, one associated with the oral apparatus, the AZM, and one a little way off, the dorsal zone of membranelles (DZM). Both are of oral origin (Corliss, 1961, p. 84). In the more advanced forms, the DZM may be more posterior or may extend so as also to encircle the anterior end. Both AZM and DZM can be retracted by the inversion of the pellicle on which they are mounted. In *Epidinium* (Sharp, 1914) the AZM surrounds the peristome which also has a ring of simple cilia around it. These are of uncertain homology.

In the odontostomes there seems to have been a reduction in both types of cilia (Fig. 5A). The AZM is reduced to 9 membranelles usually set inside the rather small buccal cavity. The somatic ciliature is asymmetrical on the two sides of the animal and is arranged in groups, many of them set on the marked ridges of the body surface.

NUCLEUS

One of the characteristic features of ciliates is that they have two types of nuclei, a macronucleus and one or more micronuclei (Fig. 6). In live material the former is usually easily visible, especially when viewed by phase contrast microscopy. The latter, however, require to be stained before they can be easily seen. Some species or strains lack micronuclei, as for instance many strains of *Tetrahymena pyriformis*. Conjugation is obviously impossible in such animals but there is no evidence that they are otherwise impaired by the deficiency.

The macronucleus is very variable in shape and size and usually stains heavily by the Feulgen method, indicating a high DNA content. The macronucleus has been shown to be, in most cases, polyploid containing many copies of each of the haploid set of chromosomes. It always arises from the micronucleus and its origin is considered in Chapter 9. Most ciliates have a single macronucleus and the commonest shape is that of a sphere or ovoid (e.g. *Paramecium, Tetrahymena*). This is almost always the case in smaller ciliates. In larger forms, however, the macronucleus may be elaborate. In *Spathidium spathula* it is very elongated, its total length being twice that of the body. In *Spirostomum ambiguum* also it is long but here there are constrictions at intervals giving it the appearance of a string of beads (moniliform). A more or less horseshoe-shaped macronucleus is found in almost all peritrichs and in *Euplotes* and *Didinium*. The most complex macronuclei are found in suctorians. *Gorgonosoma arbuscula* for instance has a much-branched nucleus echoing the branches of its tree-shaped body while *Ephelota gemmipora* has a basket-shaped macronucleus.

Some ciliates have more than one macronucleus. *Stylonichia mytilus* has two, one at each end of the body, while *Dileptus anser* has many hundreds. There does not appear to be any particular relationship between the taxonomic position of the animal and the number of macronuclei it possesses. Indeed even within a single genus the number and form may vary considerably. Thus different species of *Dileptus* have fragmented, elongated and moniliform macronuclei while in different species of *Spirostomum* they may be moniliform or ovoid.

On the whole there is a close correspondence between the size of

B

the animal and its macronucleus. The more elaborate shape of the macronucleus in large ciliates is of obvious importance in bringing all parts of the cytoplasm close to the nucleus and also in increasing its surface area for nuclear-cytoplasmic exchanges. It is worth noting, however, that whatever the shape of the nucleus, it becomes a simple ovoid or sphere at the time of division.

The macronucleus when stained shows many of the normal nuclear features: membrane, Feulgen-positive chomatin and nucleoli containing RNA. The clear areas of nucleoplasm termed karyolymph are usually difficult to see because of the great density of the chromatin. The nucleoli may vary through the animal's life-cycle and may even show changes correlated with bouts of feeding. For instance, Vivier (see Raikov, 1969) has shown in *Paramecium caudatum* that before feeding the nucleoli are dispersed whereas after feeding they congregate into groups which disappear some 9–18 hours later. In *P. bursaria* the usually large nucleoli break down into many smaller ones at the time of division and may even disappear, to reappear when fission is complete. The dissolution of the nucleii is accompanied by a disappearance of nuclear RNA. Some ciliates have many small nucleoli while others have longer polymorphous ones, although in view of the changes reported above for *Paramecium* some of these differences may be transitory. De Terra (1960) has reported that some species of *Stentor* lack nucleoli but that other possess them.

A certain number of holotrichous ciliates, chonotrichs and related gymnostomes have a macronucleus that differs from the typical pattern described above. In these the strongly Feulgen-positive chromatin is confined either to one end or to the periphery of the nucleus. These nuclei have been termed heteromerous by Fauré-Fremiet (1957). The Feulgen-negative area usually contains a single darkly staining granule, the endosome. The nucleoli are often peripheral and more common in the Feulgen-positive region.

Even more atypical are the macronuclei of a number of primitive holotrichs, including the well-known genus *Loxodes*. In these animals there are always at least two macronuclei (though there may be many) and one or more micronuclei. The macronuclei do not show the elaborate morphology of other ciliates and are small (10–15 μm diameter). They are also different in that they are stained only lightly by the Feulgen method. It has now been established (see Raikov,

Fig. 5

A, *Saprodinium dentatum*, an odontostome. **B,** *Halteria geleiana*, an oligotrich. **C,** *Entodinium caudatum*, an entodiniomorphid. **D,** *Euplotes patella*, a hypotrich. **A** and **C** after Corliss (1961); **B** and **D,** after Mackinnon and Hawes (1961).

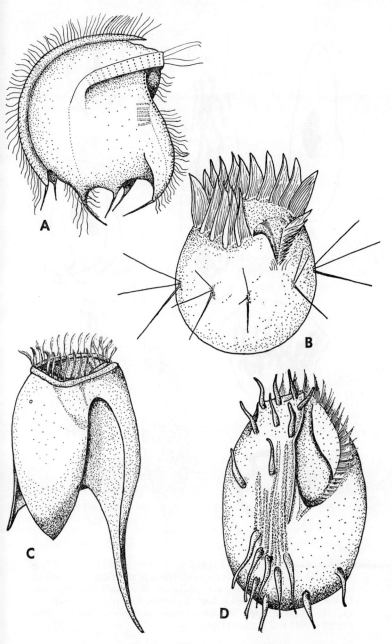

Fig. 5

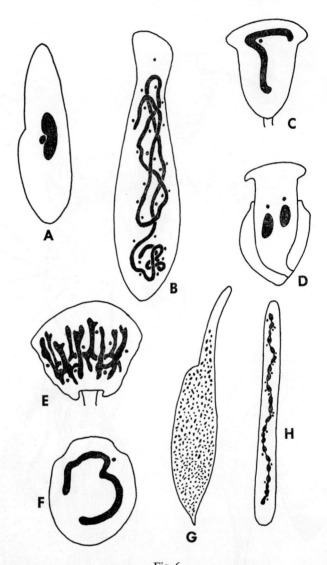

Fig. 6
The nuclei of various ciliates (both macro- and micronuclei are shown in all cases). A, *Paramecium caudatum*, B, *Spathidium spathula*, C, *Vorticella* sp., D, *Tintinnopsis* sp., E, *Ephelota gemmipara*, F, *Euplotes* sp., G, *Dileptus anser*. H, *Spirostomum ambiguum*. Partially after Corliss (1961) and Raikov (1969).

1969, for review) that in *Loxodes* (and probably in some other genera) the macronucleus is diploid, like the micronuclei. These macronuclei differ from micronuclei, however, in that they are intensively active in the production of RNA. They contain nucleoli which sometimes follow a cycle of activity suggesting the manufacture of RNA followed by its export from the nucleus. When the animal divides these diploid macronuclei do not divide but are distributed between the daughter cells (Fig. 7). Their number is made up by mitoses of

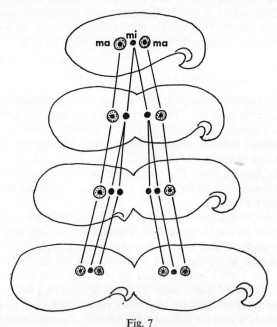

Fig. 7

Nuclear reorganization during division in *Loxodes rostrum*. The macronuclei (ma) do not divide but are supplemented at each cell division from micronuclei (mi). After Raikov (1969).

micronuclei followed by the development of some of these into macronuclei. Not only are these diploid macronuclei unable to divide but they do not synthesize DNA (Torch, 1961). The micronuclear mitoses that make up the full macronuclear complement may occur asynchronously throughout the interdivisional period (e.g. *Loxodes magnus*) or may be almost synchronous and occur at the time of cell division (e.g. *L. rostrum* and *L. strictus*). One might suppose that macronuclei created in this way would exist indefinitely but in practice this is not the case and macronuclei have been observed to degenerate, presumably to be replaced by extra micronuclear mitoses. Raikov (1969) estimates that a macronucleus in *L. magnus* will survive about 4–7 cell generations. In all other ciliates the macronucleus is polyploid, new macronuclear material being produced only when the original one has been destroyed at conjugation or autogamy.

The micronucleus is a much less variable structure. Ciliates may contain one (e.g. *Vorticella*) or many (e.g. *Spirostomum ambiguum*, about 20). The number is not constant within a genus. They are small, often only 1–2 μm in diameter, but sometimes larger, and spherical in shape. They divide mitotically, although centrioles are lacking, and during sexual reproduction undergo meiosis.

CONTRACTILE VACUOLE

One of the more obvious features of many living ciliates is the contractile vacuole (Fig. 8). This is constantly undergoing a cycle of a gradual expansion (diastole) followed by a sudden collapse (systole) as the contents are discharged to the exterior. The cycle may vary in length from a few seconds in a small active ciliate to as much as five minutes in *Spirostomum*. The contractile vacuole is usually considered to be an osmoregulatory organelle and is not found in some marine species (see Chapter 5).

In its simplest form it is spherical, discharging through a permanent pore on the body surface. The site of the pore may stain by the silver method and have a definite position (e.g. *Tetrahymena*). There may be more than one such vacuole, as in *Podophrya*, in which 1, 2 or 3 are common. New vacuoles may appear in this suctorian if it is placed under osmotic stress by lowering the concentration of the external medium. A further elaboration is to be found in those animals whose vacuoles are fed by one or more canals draining the parts of the animal distant from the vacuole. In this case, most of the fluid will be produced in or near the canals and the vacuole acts mainly as a reservoir and expulsion organelle. In *Spirostomum* the contractile vacuole is at the posterior end, discharging terminally, and is fed by a single canal running the length of the animal. If this canal is blocked, as can happen if the animal is slightly squashed under a cover slip, fluid accumulates and the canal becomes distended. In the large flat *Frontonia* the vacuole is more or less central with radiating feeder canals. *Euplotes* has a large number of fine feeder canals which drain into small vesicles which fuse to form the contractile vacuole. The fusion of contributory vesicles is a common method of contractile vacuole formation (e.g. *Blepharisma*).

One of the most complicated vacuolar systems is that found in

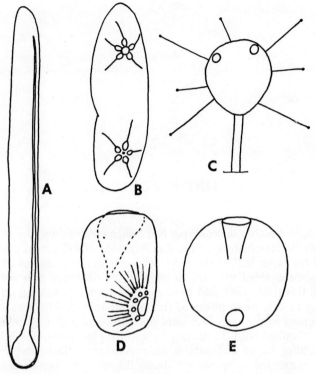

Fig. 8
The contractile vacuole systems of **A**, *Spirostomum ambiguum*.
B, *Paramecium aurelia*. **C**, *Podophrya collini*. **D**, *Euplotes patella*.
E, *Holophrya nigricans*. **D** after King (1933).

most species of *Paramecium*. There are two vacuoles discharging
through permanent pores. Each vacuole is fed by a system of radiating canals. The fluid accumulates in these canals to form swollen
ampullae which discharge their contents into the vacuole which in
turn empties to the exterior. To function efficiently this system
requires considerable coordination of the parts and great structural
complexity including valve-like mechanisms between the ampullae
and the vacuole to prevent back-flow at systole.

The vacuolar apparatus of *Haptophrya michiganensis* (MacLennan,
1944) is composed of a wide canal with regularly spaced side tubes
opening to the exterior. A systolic wave passes along the canal expressing fluid through the side tubes. Diastole involves the emptying
into the canal of contributory vesicles.

2

FINE STRUCTURE

Although electron microscopes had been available from the 1940s, a thorough examination of the fine structure of ciliates was delayed until the latter half of the 1950s when the development of ultramicrotomes enabled workers to prepare the necessary thin sections. Until this time, work had been confined to whole mounts of such things as cilia or to pieces of fragmented animal. Attempts at thin sectioning in the early '50s were not successful and no good results were obtained until the development of the glass knife, suitable hard embedding media and suitable ultramicrotomes. Since that time we have witnessed a revolution both through the proliferation of techniques and the breadth of their application and through the quality of the microscopes (Pitelka, 1970). As expected, electron microscopy passed first into a descriptive phase; a time of wonderment at the wealth of unsuspected, enthralling but little understood detail that stood revealed. Quickly, however, the discipline matured with the electron microscope being used as an experimental and comparative tool in cytological, physiological and biochemical investigations. One of the more recent developments has been that of the scanning electron microscope with which we are able to examine the surface of three-dimensional objects at high magnifications.

THE PELLICLE

Most ciliates have a very characteristic body form and there is no doubt that this is largely due to the rather stiff pellicle. Asterita and Marsland (1961) have shown that certain enzymes, especially proteases, render *Paramecium* much more susceptible to deformation and membrane rupture when subjected to high hydrostatic pressure, which presumably indicates the importance of protein in structural maintenance of the pellicle. On closer examination the pellicle is often seen to be sculptured. Thus *Paramecium* has a reticulum of rhomboids, *Stentor* and *Spirostomum* longitudinal stripes and some species of *Vorticella* horizontal stripes. In others, for example the suctorian *Podophrya*, the surface appears smooth.

The term pellicle is widely used to describe the assemblage of structures which makes up the outer covering of the animal. Where the pellicle stops and the interior or endoplasm begins is not easy to define and the distinction is often an arbitrary one. This is especially true where there are large or complex systems of sub-pellicular fibrils, and in many ciliates a thick zone or cortex is present over the surface which may include, as well as the pellicle, the ciliary basal bodies, a variety of fibre systems and mitochondria. However, despite this variation, there emerges a general pattern of pellicular ultrastructure, which appears common to all of the ciliate orders. An examination of the pellicle of *Paramecium* provides a typical example. Ehret and Powers (1959) described the surface of this animal as being organized into ciliary corpuscles. These units of surface structure appear, in reconstructions from electromicrographs, as boat-shaped depressions, from the centre of which arise one or two cilia. Subsequent work (Pitelka, 1965) filled in the details of this structure and the present interpretation can be seen in Fig. 9 (see also Plate 1). The whole surface of the animal is covered by a typical unit membrane made of three layers, an electron-transparent one sandwiched between two electron-dense ones. This is the true cell membrane and it is continuous even over the cilia. Beneath this are membrane-limited alveoli, apparently distended with fluid. Each ciliary corpuscle has two such elongate alveoli aligned parallel to the rows of cilia. The alveoli of one corpuscle are in contact with those of adjacent corpuscles and in *Paramecium* at least there are perforations connecting adjacent alveoli (Allen, 1971). Alternating with each cilium, or pair of cilia, in the row are trichocysts, and beside each cilium is a small invagination of the cell membrane called a parasomal sac. Thus almost the whole surface of the animal is covered with alveoli, interrupted only by the surface organelles just mentioned. The membranes which delimit the alveoli are also unit membranes and on the outer surface of the alveoli are clearly visible in electron micrographs. On the inner surface, however, they are in intimate contact with the cytoplasm and are often elaborated with fibrous or tubular elements to form an epiplasm.

When stained with silver another surface pattern is revealed, the so-called silver-line system (Plate 4). In some cases this coincides with surface sculpturing but not always. Because of the importance

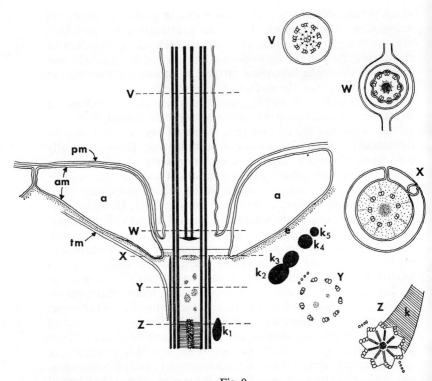

Fig. 9

Vertical section through the pellicular region of *Paramecium multi-micronucleatum* (as seen from the anterior). Cross-sections of the cilium shaft and base at the levels indicated on the vertical section (V, W, X, Y, Z). a, alveolus; am, alveolar membrane; e, epiplasm; k, kinetodesma; k_1, kinetodesma of cilium shown; k_2, kinetodesma from adjacent cilium; k_3, k_4 and k_5, kinetodesmata from other cilia; pm, plasma membrane; tm, transverse microtubules. For further details see text. After Pitelka (1965).

of the silver-line system in taxonomy a number of attempts have been made to identify, in the electron microscope, the sites of deposition of the silver. Pitelka (1961) was specifically concerned with identifying the silver-stained structures in 3 tetrahymenids. In *Tetrahymena* the silver-line pattern takes the form of longitudinal lines (primary meridia) joining the kinetosomes, which also stain, alternating with lines without kinetosomes (secondary meridia). Cross-links between meridia also occur. In fragments of sonicated animals these meridia show as well defined lines of apparently greater thickness than the surrounding pellicle. However, when the cortex is examined in section there is no sign of any thickening but it is apparent that most of the animal is covered by three membranes. As in *Paramecium* these three membranes are an outer cell membrane and the inner and outer membranes of the alveoli, which in this case are not usually distended, but flattened (Fig. 11). Deposition of silver coincided with the place where one alveolus abutted on to its neighbours, thus marking its boundaries. Deposition did not take place along cracks or ridges on the surface as had often been supposed. Further support for this theory of the site of silver deposition comes from the trichostome *Colpoda* (Rudzinska *et al.*, 1966). Here the silver pattern is confined to the kinetosomes. Electron micrographs of the pellicle show three membranes as before but while the outer two follow each other closely as in the previous examples, there is no visible connection between the two alveolar membranes except at the places at which the cilia emerge. Thus in this animal the alveolar space is continuous over the whole surface. More direct evidence was obtained by Dippell (1962) when she compared electron micrographs of silver-stained and unstained animals. She found that most of the silver was on the outside and that it was 'localized precisely at the junction point of the outermost two sets of double pellicular membranes with the innermost double membrane'. Silver was also deposited at the junction of the cilium and the kinetosome. Recently Allen (1971) has suggested that alveolar junctions may not be the sites of silver deposition. He found in *P. caudatum* that the alveolar boundaries do not always follow the cortical ridges. However, he has shown that there is granulo-fibrillar material present in the ridges which may be an alternative site of deposition.

This pellicular structure of an outer cell or plasma membrane with membrane-bounded alveoli beneath resulting in the sequence (from outside to inside) of membrane–membrane–space–membrane–cytoplasm appears to be widespread in the ciliates. It is found, often without any surface architecture, in suctorians (e.g. *Tokophrya, Podophrya*) although the ectoparasite *Phalacrocleptes* (Lom and Kozloff, 1967) has a surface covered with villus-like projections resulting from projections of the two outer membranes. At the tips of the tentacles of *Tokophyra* the pellicle is reduced to a single membrane covering the cytoplasm (Fig. 21). It is this part of the tentacle that has the ability to capture the prey and lyse its pellicle so that the prey's membranes fuse with those of the tentacle. In *Ephelota* (Rouiller *et al.*, 1956a) the alveoli are much enlarged and the pellicle has the appearance of a series of distended vesicles.

In hypotrichs also the pellicle is alveolate (Gliddon, 1966). The innermost membrane, as in other ciliates, shows elaboration into an epiplasm with a granular layer and a complex of microtubules run-

ning in various directions, connecting, in some cases, with the infraciliature. These structures presumably give the pellicle its greater
rigidity and may help to support the cirri and membranelles. As in
other ciliates the silver-line pattern probably results from deposition
along the borders of the alveoli. In the elaborate rumen ciliates, the
ophryoscolecids, there is little or no sculpturing but the surface is
very stiff and, in places, elongated into spikes giving the animal its
bizarre appearance. The pellicle has a thickness of $1 \cdot 5$ μm and appears
in the light microscope as two layers, an outer with longitudinal
fibres and an inner with transverse fibres (Noirot-Timothée, 1960).
In the electron microscope the organism is seen to be covered with the
familiar alveolate structure but with a very thick and complex
epiplasm including a layer of fibres 15 nm in diameter running longitudinally with larger transverse fibres beneath that.

The alveolate pattern is found as well in the peritrichs (Randall
and Hopkins, 1962) and here the alveoli are typically distended and
supported by ridges in the underlying cytoplasm which, coinciding
with the surface sculpturing, are easily visible in the light microscope. The tops of these ridges are dense in the electron microscope
and presumably stiffened in some way.

Although widespread the alveolate pellicle may not be ubiquitous.
In the gymnostomes *Dileptus* and *Chlamydodon* (Kaneda, 1962) the
outer membrane appears to be in direct contact with the cytoplasm,
apparently without any intervening alveoli. However, the micrographs
are not good and a thorough examination might reveal a more typical
structure. The cyrtophorine gymnostome *Nassula* certainly has an
alveolate pellicle (Tucker, 1971b). In the heterotrichs there is evidence
that alveoli may be confined to small areas of the animal. *Stentor*
and *Spirostomum* have been of constant interest to the electron
microscopist, mainly because of the contractile myonemes (see p. 54).
Micrographs show that these animals possess a cortex 10 μm or so
thick. This zone is different from the more central cytoplasm in that
it is less vacuolated and contains a greater density of organelles
organized to a greater degree. Among them are the various fibrillar
systems, and, in *Spirostomum* (Plate 2b), the fibrous layer which
separates the cortex from the endoplasm. The animals are covered
with a membrane. This is composed of two dense layers with a
transparent one between, the whole being some 50 nm thick (Finley
et al., 1964). Although the dense layers have not been further resolved, they may well prove to be unit membranes, but it is extremely
unlikely that either would prove to be two unit membranes like the
outer part of an alveolar pellicle. Underlying this pellicle is a system
of longitudinal tubules and the whole surface of the animal is
thrown into ridges and furrows about 4 or 5 μm wide and deep. How-

ever, it is interesting that in the oral region *Stentor* does show a alveolate pellicle (Randall and Jackson, 1958), suggesting that most of the surface has perhaps been derived from the more typical structure.

What is the functional significance of these pellicular structures? The major functions of the outer layers of such organisms must be the maintenance of form and the separation of the internal and external environments. How the former is fulfilled depends upon the type of animal. Some ciliates are flexible and indeed bend in their normal movements. These animals must have a pellicle which, while not being rigid, still maintains the body form more or less constant. This requirement is well met by the thin pellicle with underlying microtubules as found in *Stentor*. The ophryoscolecids and odontostomes on the other hand with their fixed shapes, often reminiscent of Roman helmets (Fig. 5A), have an elaborate epiplasm with a fibrous layer strengthened by embedded microtubules running in two directions. In forms with intermediate crustiness there is a less exaggerated elaboration of the pellicle, though it is usually the innermost layer that is thickened. In *Paramecium* the turgidity of the alveoli may help the animal to keep its shape.

The second function, of separating the cytoplasm from the external medium, depends to a large extent on the impermeability of the pellicle to a wide variety of substances. Thus for freshwater ciliates it is necessary to maintain the internal salt concentration above that of the environment, and metabolites and biologically important substances must be prevented from diffusing outwards (except in the case of waste products). Also, of course, osmotic entry of water must be kept to a minimum. However, the cell is dependent upon its environment for a variety of substances which must enter through the pellicle, oxygen and some organic compounds. Carbon dioxide and nitrogenous waste must be able to move outwards. Thus the pellicle must strike a delicate balance between permeability and impermeability and must, one suspects, exert some degree of control over many of the substances which normally pass through it. The alveolate pellicle is well adapted to form an impermeable barrier with the alveoli acting as buffer regions, with possible physiological specialization of the three membranes accounting for any selective properties.

However, some heterotrichs at least seem to be able to manage without a complete covering of this type of pellicle. Perhaps in these relatively large animals permeability is increased to allow more oxygen to reach the organism while its large volume (in relation to surface area) can cope with the increased water entry and the loss of biologically useful substances.

CILIA

One of the most striking organelles of ciliates is, of course, that from
which they take their name. From the point of view of their fine
structure cilia are both exciting and frustrating. Early workers with
the light microscope postulated the presence of fibres in cilia and the
first electron micrographs of whole mounts sometimes showed the
ends of the cilia frayed to reveal a number of fibres, usually 11, of
which 2 were different. With the advent of thin sectioning techniques
it soon became evident that this $9+2$ pattern was common to all
cilia and flagella and in the wake of this discovery came reports of
non-motile organelles that have a similar structure and are assumed
to be derived from cilia (this list includes centrioles, part of retinal
rods and insect sense organs). This extreme constancy of appearance
has resulted in most books on the subject (this one included, as can
be seen from Fig. 9) boldly advancing descriptions and pictures of 'the
typical fine structure of a cilium'. The difficulty in writing about ciliate
cilia is that some of the relevant work has been conducted on cilia
from other animals, e.g. bivalves or flagellates. However, I shall
draw on this work in some cases as there can be little doubt as to its
relevance to ciliates.

Fig. 9 shows the structure of a cilium of *Paramecium* from the
work of Pitelka (1965) and the following description refers to it for
the greater part (see also Plates 1 and 2B). The cilium can be divided
into two main regions, the basal body or kinetosome and the shaft.
The kinetosome lies below the surface and in most ciliates it is open
at the bottom. It is about $0\cdot5$ μm long and $0\cdot15$ μm in diameter. In
cross-section it can be seen to be a hollow cylinder composed of 9
sets of fibrils. Each of the 9 sets is composed of 3 tubular subfibrils,
each of which share common walls, and so cannot be entirely independent. Each subfibril is about 24 nm in diameter. These subfibrils
are inclined inwards towards the centre of the kinetosome and are
designated A, B and C, A being the innermost, most clockwise one
when seen from the inside. Fine filaments join the adjacent A and C
fibrils and also run from A spokewise to a central dense hub. This
'cartwheel' is only found in the basal third of the *Paramecium* kinetosome but its extent varies in other genera. Further up the basal body
only the fibril triplets are visible with some disorganized granular
material.

Near the level of the membrane the C fibrils taper to nothing and
the inward skewing is reduced. The top of the kinetosome is closed
with a plug of electron-dense material about 30 nm thick and above
that is a delicate septum. This complex is called the terminal plate
and marks the limit of the kinetosome. When cilia are shed, naturally

or by chemical induction, the terminal plate remains intact, and the structures above this can be considered to constitute the axoneme (Plate 1). A conspicuous feature of the axoneme is the two central fibrils, each 24 nm in diameter spaced 35 nm apart, centre to centre. They arise from a dense granule just above the terminal plate, called the axosome. Just above this the cilium presents, in cross-section, its typical appearance, most clearly shown by the Markum rotation technique (Allen, 1968). The outer membrane is continuous with the membrane that covers the whole of the animal, and inside this are the 9 pairs of peripheral fibrils. The A subfibril bears a pair of arms. Fine filaments extend from the fibril pairs towards the centre of the cilium, stopping at the central pair in *Paramecium* and at a membrane which surrounds the central pair in *Tetrahymena*. This structure, at least in motile cilia, is remarkably constant except for minor varia-tions. In non-motile cilia, typically, the central pair of fibrils is lost. There is some evidence from dense ciliary fields that the central pair are aligned at right angles to the direction of beat, but this is not a rigid rule and flagella, at least, often have the ability to beat in any direction. At the tip the cilium usually ends bluntly and arms, spokes and fibrils disappear in no fixed order (but see p. 49), although in some flagellates the central pair are prolonged well beyond the ter-mination of the peripheral fibrils. The external membrane of the flagellum is often ornamented and bewhiskered in a variety of ways but the cilium possesses almost nothing of this sort. Often it appears wrinkled in both transverse and longitudinal sections (though how much of this is due to processing artifacts is difficult to say) and occasionally there are small blobs and tubular extensions of the membrane.

In many orders cilia are combined in functional groups, mem-branellés and cirri, which act like single cilia (see Chapter 1 for gross features). In fine structure the cilia of these compound organelles appear identical to that described above. Apart from fine tubular out-pocketings which could intertwine there is little evidence of firm attachment of the cilia one with another. In fact in ophryoscolecids similar compound structures are seen to fray into their individual cilia and then reform (Noirot-Timothée, 1960), so at least here there cannot be any very permanent binding structures.

In *Euplotes* (Gliddon, 1966) the cilia of the cirri are packed hexa-gonally, the number varying both in the same cirrus in different animals and between different cirri on the same animal. The anal cirrus has 87–120 and the caudal cirri 40–46. The ground plan of most of the cirri is more or less circular. The same animal also possesses membranelles and in this case the plan is rectangular, being made up of 2 or 3 parallel rows of cilia with perhaps as many as 40 per row.

Although Gliddon could see no structures holding the component cilia together he noticed vesicles opening on to the surface of the animal close to the bases of the cirri and membranelles and suggests that they may secrete a substance which could stick the cilia one to another. Many ciliates have similar vesicles called parasomal sacs. In *Paramecium* these structures appear as small finger-like inpushings of the cell membrane just to the right of each cilium. In *Tetrahymena* sacs not unlike parasomal sacs, called mucocysts, are to be found on the primary meridia alternating with the cilia, and also on the secondary meridia (Fig. 11). When the membrane over the mucocyst ruptures the contents expand and are expelled and apparently form a slime coat around the animal. Satir *et al.* (1972) suggest that when the mucocyst reaches the surface of the cell the membranes suddenly rearrange themselves to open a continuous hydrophilic canal from the outside of the cell to the inside of the mucocyst. The concentrated material within would then draw in water by osmosis, swell rapidly and expulsion of the hydrated material would result.

The kinetosomes of membranelles, in contrast to the ciliary shafts of *Euplotes*, showed very complex interconnections. In both cirri and membranelles there are fibrous linkages along all three axes of the hexagonal array at the very base of kinetosomes. Half way up the kinetosomes is another set of fibrous connections, but these are different in cirri and membranelles. In the former they run along only one of the axes of symmetry, giving off short branches which attach to the kinetosomes. In the latter, fibrous material surrounds the kinetosomes and appears to connect each basal body with four others. Membranelles are of course found in other ciliates and have been investigated especially in the heterotrichs, *Spirostomum* (Finley *et al.*, 1964), *Stentor* (Randall and Jackson, 1958) and *Blepharisma* (Kennedy, 1965). They have a similar structure to those from *Euplotes,* but have not been investigated in such fine detail.

The electron micrograph studies of cilia discussed so far have done little more than describe the fine structure; the functional aspects have been largely ignored. At first workers were fascinated by the extreme constancy of the $9+2$ plan and for a while thoughts centred around establishing a logical reason for this arrangement. A variety of ideas were put forward, none very satisfactory though some richly imaginative (e.g. Harris, 1961). Later, however, attention turned from this still unsolved problem to that of assigning functions to the various parts; which of them is contractile? What are the antagonists to the contraction? In this field two pieces of work have changed our whole outlook on ciliary physiology and fine structure and both illustrate the new role of the electron microscope as a tool in investigations not strictly morphological.

The first example came from Gibbons (1963, 1965). He was concerned with breaking down the cilia of *Tetrahymena* into the constituent elements for analysis. Up until that time, although cilia could be isolated in quantity and reactivated by ATP, it had proved impossible to obtain any of their proteins in solution. Gibbons showed that if the outer membranes of the cilia were removed with digitonin the remaining fibrillar elements were easily attacked. Of the total protein about 36 per cent was soluble in digitonin and the remaining insoluble fraction constituted about 50 per cent of the original protein but retained almost all the ATPase activity. Examination in the electron microscope showed this fraction to contain cilia, complete and intact except for the membranous elements. Further treatment of this fraction with a chelating agent at low ionic strength resulted in all the ATPase activity coming into solution together with 30 per cent of the protein. Examination of the remaining insoluble fraction showed outer fibre pairs only; the central pair and the arms of subfibril A had disappeared. Curiously, the peripheral pairs remained with the arrangement that they would have in intact cilia and the cylinders were only occasionally broken open. Presumably their relative positions were maintained by largely invisible structures. By raising the magnesium concentration of this preparation Gibbons was able to reconstitute the cilia to some extent. In these reconstituted preparations over half the ATPase activity returned to the insoluble fraction and when he examined this in the electron microscope the arms of the A subfibril had returned 'with a remarkable degree of precision to the same position they had in the intact cilia'. These experiments in Gibbons' own words suggest 'that the arms are the site of at least part of the fibrillar ATPase activity of the cilium'. (Further implications of this finding are discussed in Chapter 3.)

The second electron microscope study which threw light on the functioning of cilia was that of Satir (1965). The work itself was carried out on the cilia from the gills of freshwater mussels but is relevant here because of its fundamental connotations and because gradually it is being confirmed for ciliate cilia. Satir examined carefully the order in which the peripheral fibres terminate as one approaches the tip of a bent cilium. Before we look at his results let us examine the possible theories of ciliary contraction assuming that the peripheral fibres are the active elements (Fig. 10). The three alternative modes of action are (a) the fibres on one side could shorten thus pulling the cilium over; (b) the fibres might not shorten but move relative to one another, sliding past each other by a mechanism similar to that proposed for muscle contraction; or (c) some of the fibres could 'wrinkle'. As Satir found no evidence for the

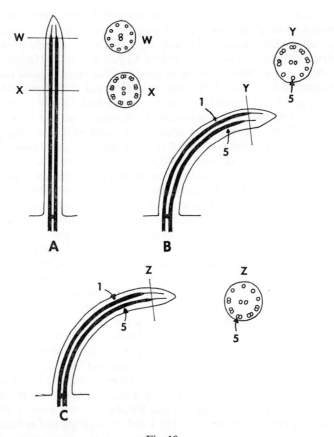

Fig. 10
Possible arrangements of microtubules in an actively bending cilium.
Longitudinal sections and transverse sections at levels indicated.
A, unbent cilium. **B,** cilium bent assuming microtubular shortening.
C, cilium bent assuming no change in length of microtubules. For
further explanation see text. After Satir (1965).

last alternative he only considered the first two. Fig. 10 shows the
expected arrangement of the fibre pairs in sections of cilia according
to the two theories. On the assumption that fibre pair 5 contracts,
transverse sections of the tips should show either disappearance of
all the pairs at once or 5 first and 1 last. If however the fibre pairs
were unchanged in length and the movement only relative then 5
would protrude beyond the others and be the last to disappear. This
latter is in fact what Satir discovered. The order of disappearance

was 9, 1, 8, 2, 7, 6, 3, 4, 5 and strongly supports the sliding filament hypothesis, of which more later (see Chapter 3).

INFRACILIATURE

Although we are now beginning to understand the biochemical workings of the cilium we still do not fully understand the factors which control the coordination of fields of cilia. This coordination, so obviously present in metachronal movement and in the 'crawling' of hypotrichs, has held out a challenge to protozoologists for over a century. Until recently workers have sought some neuroid or quasi-nervous system in or under the animal's surface. The most likely-looking structures for this function have always been the kineto-desmata. These are fine fibres that run along the longitudinal rows of cilia and apparently connect lines of basal bodies. The complex of a row of cilia and its associated kinetodesmata and kinetosomes is known as a kinety. The kinetodesmata always run on the animal's right-hand side of the basal bodies (the so-called 'rule of desmo-dexy'). As can be seen from the discussion of ciliary movement (Chapter 3) the role of these fibres is not yet fully understood but our concern here is with their fine structure.

In *Paramecium* (Pitelka, 1965) the kinetodesma is broadly attached to the anterior side of the lower third of the kinetosome and run first slightly to the animal's right then forwards. It joins the fibrils coming from more posterior kinetosomes and gradually tapers anteriorly. Transverse sections show about 4 or 5 kinetodesmal fibrils (see Fig. 9) indicating that they peter out after covering the distance between 4 or 5 cilia. The fibrils rise towards the surface of the animal as they run forwards so that in section the nearest arising fibril will be thickest with 3 or 4 more getting smaller above it. The fibrils do not appear, in cross-section, to have any very ordered sub-structure. Negatively stained whole mounts, however, show that the fibrils are cross-striated along their length in a complex way. The main repeating period is 30–35 nm but a variety of sub-bands exist. Similar fibrils have been found in other ciliates.

There are many other fibrillar and microtubular systems associated with the kinetosomes, and it has been possible over the years to obtain large quantities of comparative data (Pitelka, 1969). Out of this data it has been possible to arrive at consistent nomenclature of parts and the recognition of homologies (see Grain, 1969, for review). A fairly generalized arrangement of the infraciliature can be seen in *Tetrahymena* (Allen, 1967, Fig. 11). From the kinetosome the kineto-desma arises and runs towards the anterior and on the animal's right. Two sets of microtubules arise near the basal body. The transverse set

arise on the anterior side of the kinetosome and pass upward towards
the surface and to the animal's left terminating close to the adjacent
kinety. The post ciliary microtubules arise at the right posterior side
of the kinetosome and run obliquely towards the posterior passing
over the kinetodesmata. In this animal there are also two other sets
of microtubules running longitudinally that do not appear to be ciliary
derivatives. These are the deep basal microtubules and the more super-
ficial longitudinal microtubules. The arrangement of the infraciliature in *Paramecium caudatum* is essentially similar (Allen, 1971).
In addition to the two sets of ciliary-derived microtubules there are
also two fibrous networks, a deep infraciliary lattice and the super-
ficial 'striated bands' composed of microfibrils and probably contractile. There are various theories as to the functions of these
ordered and complex arrangements of tubules and fibres. The most
likely ideas are that the three elements originating from the kinetosome anchor the cilium during its movements. These and the other
systems are also probably responsible at least in part for the maintenance of the spacing between the cilia and the kineties. The striking
microtubular structures of *Stentor* and *Spirostomum* (Plate 2B, Fig.
12), equated with kinetodesmata and called km fibres by Randall
(1957), have been shown to be homologous with the postciliary microtubules (Bannister and Tatchell, 1968; Grain, 1968). In cross-section
the km fibre is composed of 20–22 nm tubules arranged in layers
stacked one upon the other. The number of layers varies and may be
as many as 66. Each layer has between 18–22 microtubules with
fine connections between each, and is derived from a single kinetosome. There are interconnecting fibrils from one layer to another
more or less at random, but the third microtubule from the outside
invariably has such a connection As the animal is contractile these
sheets of microtubules must be able to slide over each other and at
such times these connections between layers must be broken and
remade. Similar km-like structures are to be found in other heterotrichs (Kennedy, 1965).

Fig. 11

Three-dimensional reconstruction of a small portion of the cortical
region of *Tetrahymena pyriformis* showing part of two kineties and the
area between. The anterior of the animal is to the top and the animal's
left is to the reader's right. Note especially the three tubular and fibrous
elements arising in the close proximity of each kinetosome, namely the
kinetodesma (k), the transverse microtubules (tm) and the post-ciliary
microtubules (pm). The longitudinal microtubules (lm) and the basal
microtubules (bm) do not appear to arise from or near kinetosomes.
a, alveolus; c, cilium; e, epiplasm; m, mitochondrion; mb,
mucigenic body (mucocyst). After Allen (1967).

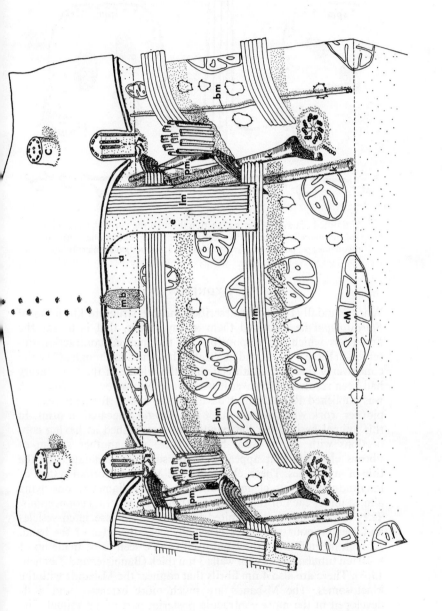

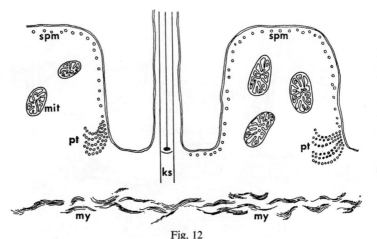

Fig. 12

Simplified diagram of vertical section through the corticular region of *Spirostomum ambiguum*. ks, kinetosome; mit, mitochondrion; my, myoneme; pt, posterior microtubules; spm, subpellicular microtubules.

MYONEMES

When Randall (1957) first described the fine structure km fibres of *Spirostomum* he identified them as myonemes, that is to say the organelles which are responsible for the marked contraction observed in these animals and in *Stentor*. It seemed as though at last the controversy that had continued since the middle of the last century had been resolved. However it was not to be, and even as this work was published Randall must have had his doubts for next year his pioneer work with Jackson (1958) on *Stentor* appeared in print. In this animal besides the km bands, now established as having connections with ciliary basal bodies, there was another system of fibrous elements that appeared to be suitable candidates for the title of myonemes. These, christened the M-bands, form a system of longitudinal fibrous elements about 1 μm in diameter, one lying almost under each of the kineties with occasional cross connections. They are not membrane limited but often have associated vesicles and membranous structures especially near the edge of the band. At high magnification the band itself can be seen to be made up of 8–10 nm tubular fibrils with walls 3 nm thick (Bannister and Tatchell, 1968). There are also 4 nm fibrils that connect the M-bands with the kinetosomes. The M-bands are much more extensive and well-developed in the more contractile posterior part of the animal. The position was now somewhat confused; *Spirostomum* contracted but had no M-bands, *Stentor* contracted and had both km fibres and M-bands. Randall and Jackson credited both systems with some contractile ability and there the matter rested until the non-contractile *Blepharisma* was shown to have km fibres. Perhaps as Pitelka (1963) said, '*Blepharisma* is contractile and doesn't know it.' Later however *Spirostomum* was shown to have a network of fibrous material rather like M-bands, and in the same position, but more diffuse (Plate 2B and Fig. 12). The fibrils of this network are only 3–4 nm in diameter and the system bears no relationship to the kineties and corticular furrows. As with the M-bands there are many associated vacuoles. *Blepharisma* as expected does not appear to possess this M-band-like system, although there is some ambiguous evidence. On this basis then the contractile properties of the microtubular elements fell into disrepute until reinstated in a new guise by recent work on *Stentor* (see p. 86). We seem to be only a little nearer to a solution of this problem than we were in 1957 but the position is not perhaps as depressing as Finley *et al.* (1964) would claim when they wrote that 'the magnitude of the advance is almost infinitesimal in comparison to the progress yet to be achieved in order to develop a logical valid comparative ultra-anatomy of heterotrichs'. Indeed I would claim that heterotrichs have proved most useful material in comparative work with ciliates.

The contractile material in peritrichs was identified earlier. Those genera that have contractile stalks (e.g. *Vorticella*) possess in their stalks a prominent myoneme that can be easily seen with the light microscope (Fig. 13A). Non-contractile genera (e.g. *Campanella*) have stalks without myonemes (Fig. 13B). The stalk of the contractile *Vorticella* when seen in cross-section in the electron microscope has a characteristic appearance (Randall and Hopkins, 1962; Amos, 1972) (see Fig. 13). Inside the stalk, to one side only, are 15–20 fibrils some 100 nm in diameter which may be attached to the pellicle. Inside the stalk is filled, except for the central portion, with a transparent matrix containing a few 4–8 nm fibrils. In the centre, or more usually rather to the side opposite the 100 nm fibrils, is the myoneme. This is composed of microfibrils 2–4 nm in diameter. It also contains tubules some 30–100 nm in diameter running longitudinally. This myoneme is continuous with those of the zooid. In the zooid the myonemes are close to the surface, pressed against the epiplasm and with endoplasmic reticulum (ER) in close contact with its inner surface. The regions of contact between the ER and the myoneme contain complicated structures termed linkage complexes by Allen (1973) spaced out about every 1 μm along the myonemes. These may act to anchor the ER to the myoneme. It has also been suggested that this ER and the tubules act as a store from which calcium

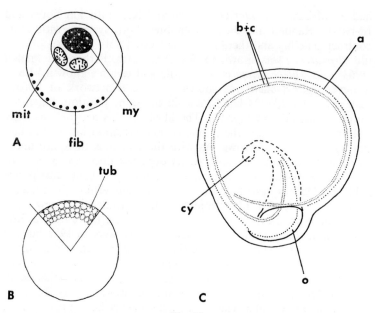

Fig. 13

Diagrams of transverse sections through the stalks of the peritrichs *Vorticella* (A) and *Campanella* (B). C, diagram showing the arrangement of the oral cilia as seen from above. a, outer peristomal cilia; b and c, inner peristomal cilia; cy, cytostome; fib, fibrils; mit, mitochondrion; my, myoneme; o, opening of the infundibulum; tub, tubule. Modified after Randall and Hopkins (1962).

may be released at the initiation of contraction rather like the sarcoplasmic reticulum of striated muscle. This analogy was strengthened when calcium was shown to be concentrated in the tubules of the stalk myoneme by Carasso and Favard (1966).

The annuli of the stalks of *Vorticella* and *Carchesium* contain 15–20 100 nm fibrils but these are not always found in other genera of peritrich. In *Zoothamnium,* a contractile colonial genus, the annulus contains hundreds of tubular elements (Plate 2A). These are about 140 nm in diameter and are cross-striated with a period of about 30–70 nm produced by the overlapping of the 6 nm fibrils that make up the walls of the tubules. Rouiller *et al.* (1956b) have elegantly shown that these tubules are derived from ciliary origins. The stalk itself is secreted by an area called the scopula, long known to contain specialized cilia. In sections passing through the scopula and showing the stalk/zooid junction one can clearly see kinetosomes in the zooid apparently connected to or secreting the tubules of the stalk. It appears that the tubule is derived from the nine pairs of peripheral fibrils. Similar structures exist in the stalks of non-contractile genera and here too they are derived from cilia (Fauré-Fremiet *et al.*, 1962). In *Opercularia* they are modified ciliary fibrils as in *Zoothamnium* but in *Campanella* they result from the elaboration of the ciliary membrane. The fibrils in the annulus of the *Vorticella* stalk are also of ciliary origin. It is assumed that these fibrils and tubules are of skeletal function acting in a supporting way in the non-contractile genera and also as a means of supplying the elasticity for extension of the contractile stalks. The asymmetrical distribution of the fibrils in *Vorticella* and *Carchesium* may play some part in ensuring a helical contraction.

THE NUCLEUS

Most of the structures with which we have so far dealt are organelles peculiar to ciliates but these animals contain many of the features and organelles of other types of cells; most obviously they contain one or more nuclei. It is a characteristic of the ciliates that they have two sorts of nuclei, micro- and macronuclei. Most of the electron microscope examinations have been of macronuclei. The nuclear membrane is in fact composed of two unit membranes, as in other eukaryotic cells. The inner one bounds the nucleus while the outer is continuous with some parts at least of the endoplasmic reticulum. The whole system is some 20 nm or more across. Characteristically it has pores in it about 50 nm in diameter and occasionally with some sort of elaboration round the edge and often with a delicate septum or plug. An interesting study of macronuclear pores in *Tetrahymena* has recently been carried out (Wunderlich and Franke, 1968). It appears that there are differences between the nuclear membranes of rapidly dividing and non-dividing animals. Thus in organisms from cultures showing no growth the nuclear membranes have simple round or polygonal pores with a 4–6 nm wide perimeter. The pores themselves are 56–65 nm in diameter and occur in densities of from 75 to 145 per square micrometre (which means that about 31 per cent of the membrane is pore!). In rapidly dividing, cells however, some had pores of this sort while others had pores with a wide perimeter, leaving a central aperture of only some 20 nm. The latter type of pore appears to be filled with some sort of material and there is a central granule 8–18 nm across. In all nuclei, however, the area of the membrane occupied by pores was very close to 31 per cent despite differences in type and frequency. This association of the latter type of pores with active cells has been found elsewhere and adds strength to the idea that nucleocytoplasmic interactions are conducted through and controlled by the pores. Wunderlich (1969) has suggested that the central granule in these pores may represent a ribonucleoprotein particle in transit from the nucleus, where it has been assembled, to the cytoplasm. The membrane of the micronucleus may not possess pores. Certainly this seems to be the case in both *Paramecium aurelia* (Jurand *et al.*, 1962) and *Blepharisma intermedia* and of course makes sense if one assumes that the macronucleus is the somatic one from which instructions for protein synthesis are sent while the micronucleus is a genetic 'memory store' not directly involved in the day-to-day workings of the cell.

In some protozoa there is associated with the nuclear membrane a cortex, usually with a honeycomb-like structure, which may have the function of maintaining the shape of the nucleus. This is especially pronounced in *Amoeba proteus* where the constantly streaming cytoplasm would tend to produce distortion, but another large amoeba, *Pelomyxa,* does not have it. The macronucleus of many ciliates has a complex form but even in these no elaboration of the membrane has been reported, although Roth (1956) found microtubules of unknown function close to the micronuclei of *Euplotes*.

Many workers have expressed their disappointment over electron microscope studies of the nucleus. At first it was hoped we might gain some insight into the fine structure of the chromosomes and perhaps into the molecular organization of the materials within them, but it has not proved to be the case. While electron microscopy has produced invaluable results correlating ultrastructure with biochemical function in almost every cellular organelle or process investigates, the nucleus when opened proved to be an almost empty box. In *Colpoda* (Rudzinska *et al.,* 1966), for instance, the macronucleus, inside its membrane, has a ground substance that is fibrogranular with no apparent order. Set in this matrix are randomly distributed concentrations of dense fibrous and granular material about 2 nm in diameter, which are taken to represent the Feulgenpositive material. Very little of any significance can be gleaned from this eminently uninteresting object. The micronucleus is much the same except that it is usually only 1–2 μm in diameter and the chromatin somewhat more densely concentrated.

As mentioned above, the macronucleus is the somatic organelle while the micronucleus retains the genetic information intact. Although some species have amicronucleate strains and so can exist without periodic renewal of the somatic DNA, there is evidence that the macronuclear material does require occasional reorganization and repair. Also of course the genetic material of the macronucleus must be doubled before division. As the macronucleus is thought to be usually grossly polyploid the simple nipping into two halves at division suffices to supply each daughter with a full complement of genes many times over (but see p. 35). The necessary increase in DNA content appears at least in *Euplotes* to be a very orderly process. The macronucleus is C-shaped and normally stains moderately with Feulgen. Some time before division non-staining bands appear at both ends of the C and progress towards the centre where they meet and disappear. That part of the nucleus which has been traversed by such a 'reorganization band' stains more intensely and appears coarsely granular. Gall (1959) showed that DNA synthesis occurs just behind the band for a short period of time and his data were consistent with the theory that this synthesis represents the doubling of a large number of small units which he assumes are the individual chromosomes. This phenomenon has been examined in the electron microscope (Fauré-Fremiet *et al.*, 1957) and here four areas were clearly defined. The first was the macronucleus unaffected by the band, which showed 1 μm electron-dense granules corresponding to the Feulgen-positive grains, typical macronuclear fine structure and much the same in appearance to the part of the nucleus through which the band had passed. The band itself had two zones, the anterior advancing side of the band is composed of fine 10 μm twisted or spiral filaments which were taken to be threads of genetic material, perhaps combined into multimolecular units. It is here presumably that DNA replication is taking place, for behind this is a relatively clear zone in which condensation occurs to reform the typical chromatin aggregates. Once again one is struck by the simplicity of the nuclear structures seen in the electron microscope when compared with light microscope studies, for Gall supposes that there is endomitotic activity at the reorganization band and yet in the electron microscope the appearance is more of a dispersal and reaggregation of granules and filaments.

In suctorians the nutritional state of the animal may affect that fine structure of its nucleus. Thus in *Tokophyra* (Rudzinska, 1956) that are either old or over-fed, some of the chromatin bodies of the macronucleus appear large and hollow and on closer examination show an ordered, almost crystalline, structure of a honeycomb lattice with cells 47 μm across and walls 12 μm thick.

MITOCHONDRIA

The mitochondria of ciliates are in no way especially remarkable. As with their metazoan equivalents they appear to be composed of two membranous sacs one inside the other, the inner one having numerous finger-like inpushings, called cristae. Protozoan mitochondria differ from those of metazoans in that the cristae are tubular rather than platelike, although occasionally the latter are found in Protozoa. It is

assumed that mitochondria in ciliates fulfil the same role as in higher
animals, that is the production of energy. Generally speaking mitochondria are found either near the oxygen supply or near a site of
energy consumption. Thus in *Spirostomum* the mitochondria are
mostly found in the 5 μm or so just under the pellicle and in rapidly
growing *Tetrahymena* they are found flattened against the surface
of the animal (Elliott and Bak, 1964b). They are found clustered
around such sites of activity as the contractile vacuole and myonemes.

In *Tetrahymena* the effects of ageing have been noticed in the
structure of the mitochondria. As the culture reaches the stage when
division ceases the mitochondria move away from the animal's
surface and become more rounded. Later electron-dense granules
appear which grow until the whole of the mitochondrial structure is
obscured and the cytoplasm dotted with these inclusions. They may
be of lipid but ageing *Tetrahymena* also produce granules of calcium
pyrophosphate, the so-called volutin granules, and the mitochondria
may well be the source of these.

CALCIFIED STRUCTURES

Many other ciliates produce calcium depositions, some with ageing,
some as a normal part of their metabolism. The gymnostome *Coleps*
(Fauré-Fremiet *et al.,* 1968) has a calcium phosphocarbonate skeleton
forming a lattice of longitudinal and transverse bars which is secreted
into the alveoli of the pellicle and presumably acts to hold the animal
firmly in shape. When the animal divides each daughter receives
half the skeleton but has to manufacture the other half. This process starts with an amorphous calcium deposition in the alveoli.
A rather dense fibrous substance then seems to be secreted into the
alveoli which 'shapes' the bars of the lattice into their characteristic
shape which appears in cross-section to be both complicated and
constant. The calcium is deposited on to a polysaccharide matrix and
mitochondria, granular endoplasmic reticulum and perhaps microtubules are involved in the process. The marine gymnostome
Prorodon (André and Fauré-Fremiet, 1967) has calcium granules
scattered throughout the cytoplasm which when examined in the
electron microscope have a concentric layered appearance. In this
case the concretions appear first in membranous sacs associated
with granular endoplasmic reticulum. These granules are similar in
size and distribution to the volutin granules of *Tetrahymena*.

In *Spirostomum* granules of the calcium phosphate bone salt,
apatite, accumulate as the animal ages (Pautard, 1959; Jones, 1967).
They are about 1 μm in diameter and composed of many needle-
like crystals about 5 nm in diameter which give a spiky appearance.

Pautard (1970) has followed their formation in the electron microscope and suggests the following sequence of events. Small vesicles appear in the endoplasm and these form by coalescence larger vacuoles. In these appear 'particle embryos' (usually more than one per vacuole) which lie close to the vacuole wall. Many mitochondria in the cytoplasm are situated close against the outer surface of this wall. The embryos gradually become calcified and, when they are mature, become separated into small individual vesicles lying in the endoplasm. The function of these and other calcium deposits is not fully understood, but they may be connected with ageing processes (Jones, 1969) or possibly with contraction (Vivier et al., 1969).

Skeletal processes not made of calcium salts are also known in the ciliates, indeed many of the microtubular elements already described in connection with the pellicle, infraciliature and stalk may have s keletal functions.

THE GOLGI APPARATUS

In metazoan cells this system is composed of many flattened saccules arranged one upon another. The saccules are apparently formed at one end of the complex from smooth endoplasmic reticulum, while at the other end small vesicles break away into the cytoplasm. The apparatus is found most commonly in cells rapidly manufacturing proteins for export and it is thought that the protein is made at the ribosomes of the rough endoplasmic reticulum (ER), passed to the space between the ER membranes and then transported to the smooth ER. This then is organized into a Golgi apparatus in which the secretion is concentrated and finally packaged into more or less spherical vesicles.

In ciliates the Golgi apparatus is not often observed, in contrast with other groups of Protozoa where it may be a common feature. It has been observed in something like the metazoan form in *Blepharisma* where it is often associated with rough ER (Kennedy, 1965); in peritrichs it is involved in the digestive process of the food vacuoles (Carasso et al., 1964). Estévé (1972) has described Golgi complexes in *Coleps, Colpidium, Stylonychia* and *Paramecium* and suggests that they are the sites of production and packaging of digestive enzymes and that the vesicles produced are, in fact, primary lysosomes. In *Tetrahymena* the Golgi apparatus is absent except at the time of conjugation (Elliot and Zeig, 1968). In the growth phase these animals have separate flattened saccules occurring in the oral region. If however the cells are starved for 12–38 hours the saccules become aligned into stacks but they do not appear to produce vesicles. When such cells are mixed with similarly treated animals of a compatible mating type conjugation occurs, and within minutes the Golgi apparatus produces vesicles. However, this activity wanes to a minimum 5 hours after the coming together of the conjugants and then increases again until at 10 hours vesicles are present once more. Exconjugants have a Golgi apparatus composed of concentric whorls of saccules, which disappear after a period of feeding and division. Presumably some substance is being produced and packaged during the two active periods but its nature is unknown. It appears not to be a hydrolase but could perhaps be one or more proteins concerned with the conjugation cycle.

ENDOPLASMIC RETICULUM

This membrane system occurring either as rough ER, that is with ribosomes, or as smooth ER is found commonly in ciliates and is presumably fulfilling the same sorts of functions as in metazoan cells. Its main function would be, therefore, the manufacture of proteins in the ribosomes. It is often found to a greater extent in places where some sort of synthesis is occurring, as with the skeleton of *Coleps*. It also seems to be involved with the process of osmoregulation; the cytoplasm around the contractile vacuoles of many ciliates is rich in smooth ER which appears to communicate with the lumen of the vacuole (peritrichs) or with the feeder canals (*Paramecium*, Schneider, 1960) (Fig. 28). There is evidence that the fluid of the vacuole comes from the ER space and it may be here that it is made hypotonic to the cytoplasm.

TRICHOCYSTS

Many, but not all, holotrich ciliates possess trichocysts. These are of two main types, non-toxic and toxic. The former are better known as they are found in *Paramecium* and will be described first. In *P. caudatum* they are about 3·7 μm long. Their structure in the undischarged state is very complex and Bannister (1972) has described some 10 different parts including sheaths, tip and contents. The contents, thought by earlier workers to be structureless, now appear to be crystalline, composed of an interwoven complex of fine filaments. Upon suitable stimulation these trichocysts discharge their contents as a thread some 6 to 8 times as long as the undischarged organelle. These discharged threads have a marked cross-striation with a 60 nm period. Bannister suggests that slight rise in pressure within the cytoplasm could cause the trichocyst tips to break through the overlying pellicle thus allowing water to enter the trichocyst bodies. This in turn results in elongation of the thread, a process which may result from a simple rearrangement of the constituent filaments from one crystalline pattern to another.

A number of functions for these organelles have been postulated, including protection, attachment to the substrate and osmoregulation. However, there is little doubt as to the function of the toxic trichocysts of gymnostomes. These are stimulated to discharge by contact with the prey organism. The victim is immobilized and in some cases partially cytolyzed. These trichocysts appear to be preformed hollow tubes that evert on discharge (Dragesco, 1952). They are unarmed and unstriated.

THE FUTURE

The foregoing brief review of ciliate fine structure shows that despite the great wealth of organelles to be found, there are grounds for optimism in those fields dealing with the description and cataloguing of static structures. More and more, however, electron microscopists are using their apparatus and techniques to investigate the life processes and dynamic changes in the organelles. In this way we have seen dramatic advances in ciliate physiology in the last decade. Simple comparative work has helped us to assign functions to various organelles but there is a limit to the information that can be gained from these sources.

Important advances have been made by the application of the techniques of light histochemistry to the electron microscope. These methods are of three main types. Firstly we have the tests for specific substances and enzymes. For use with the electron microscope the stains used must be electron dense, and heavy metals are widely used. Tests already employing these metals, such as those for alkaline and acid phosphatase (see p. 99), can be used in a similar way to orthodox histochemistry. In some cases the electron-dense material must be added as an extra. For example, antibodies labelled with the iron compound ferritin can be visualized in the electron microscope and result in the identification of the site of the antigen to which they bind (Sinden, 1971). The second type of test that has been adapted is that of selective digestion by which the nature of a structure can be elucidated by noting its removal by specific enzymes or by testing the solubilized fraction. These techniques have proved their worth in Gibbons's hands and no doubt will be used more in the future. The third technique is autoradiography, by which the site of radioactive substances can be detected. There are certain limitations imposed when this technique is used in conjunction with the electron microscope. It is only possible to use low energy isotopes such as ^3H, ^{14}C and ^{35}S and the resolution is limited by the relatively thick layer of

64 The Ciliates

film and by the grain size. Nonetheless this is a technique with great potentialities which has been somewhat neglected by those studying ciliates.

We now have a variety of tools related to the electron microscope. The most potentially useful in the immediate future is the scanning electron microscope, with which we can view at great magnifications, and with large depth of focus, the surface details of small objects. The resolution obtainable to date is about 20 nm but this will no doubt be improved before long. One limitation of this machine is that the specimen must be dried before examination. Until lately this restricted its use to hard objects such as diatom tests. Recently, however, Small and Marszalek (1969) and others (Boyde and Barber, 1969) have developed and improved the drying techniques and examination of all sorts of ciliate material has become possible. Especially impressive are the pictures of feeding in *Didinium* (Plate 3) (Wessenburg and Antipa, 1970).

Another related machine is the electron probe analyser with which it is possible to detect the presence of a variety of types of atom in thin sections of material. Already this device has proved useful in the analysis of bone and sponge spicules, and has recently been used with ciliate material (Vivier *et al.,* 1969; Zagon *et al.,* 1970).

Much has been omitted in this brief survey. However, where it has has seemed more appropriate the contribution of the electron microscope has been dealt with under other headings.

3

LOCOMOTION

Typically ciliates swim through the water with one end, usually the anterior, leading. As they move they may rotate on the anterior–posterior axis and the path that they follow may be a shallow spiral rather than a straight line. The spiralling seems usually to be produced not by the shape of the animal but by the action of the cilia (Seravin, 1970). The animal usually continues on a steady course until it meets some form of obstruction. The obstruction may be physical or it may be chemical (e.g. a region of high salt concentration). Upon meeting such an obstruction many ciliates perform an avoiding reaction in which they reverse the beat of their cilia, swim backwards a short distance, then recommence forward swimming in a slightly different direction. This manœuvre they repeat until the obstacle is skirted.

Many ciliates diverge from this typical pattern. The oligotrich *Halteria* (Fig. 5B) hangs motionless in the water then suddenly jumps a distance of many microns. The large *Spirostomum* is able to swim on a sinuous path because it steers itself by curving its body, 'worming' its way along. Perhaps most remarkable are the hypotrichs such as *Euplotes* (Fig. 5D) which frequently walk rather than swim. For this they use their ventrally placed cirri, which act like many little legs. Other ciliates may use their cilia not to move themselves but to create water currents from which they can extract food. This is the case in peritrichs and *Stentor* when attached. Such feeding currents may be created by the powerful compact membranelles of the AZM.

Some ciliates are not only capable of swimming but also have a system of myonemes whose contraction can cause considerable changes in the animal's shape. Such myonemes are found in the

c

stalks of peritrichs where their function is to retract the zooid down to the substrate, probably helping to protect it from damage. In the free-swimming *Spirostomum* the action of the myonemes may help the animal to burrow in the substrate (Pautard, 1970).

Ciliates are able by means of their cilia to swim, walk and create water currents. Cilia from different species, even different phyla, are remarkably similar in fine structure (see Chapter 2) and the wide range of functions that they perform reflect differences in their arrangement and the form of their beat rather than in their structure. Something has been said (Chapter 1) concerning the arrangement of cilia and now we must consider the form of the beat. Until relatively recently cilia and flagella were considered to be distinctly different organelles with their own characteristic type of beat. Typically flagella are borne singly or in small numbers whereas cilia are numerous, often covering the animal. The beat of flagella was thought to be sinusoidal with the wavelength shorter than the organelle as can be seen for instance in bull sperm tails. Ciliary beat on the other hand was thought to be asymmetrical and with a wavelength longer than the organelle. Most workers consider that the form of beat in all cases was like that of the abfrontal cilia of *Mytilus* gill filmed by Gray in 1930. Although Lowndes (1943) showed that not all flagella beat alike, it was not until the work of Jahn and co-workers (see review Jahn and Bovee, 1964) that it was realized that flagellar beat was very variable indeed. *Peranema* for instance can beat its flagellum in three quite distinct ways. In *Chlamydomonas* the two flagella are used in a breast-stroke-like manner, held straight out in front then swept backwards, bending only at the base. When this power stroke is finished, the bend moves along the flagella drawing them back to the start of the power stroke again. This is of course essentially the type of beat described for cilia. On the basis of this and structural evidence we can no longer consider cilia and flagella different sorts of organelle but rather ends of a spectrum of adaptations from one basic type.

The realization of differences in the ciliary beat of different species has been only fully reported in the last few years, mainly due to the work of such researchers as Párducz (1966) and Sleigh (1962, 1968). The latter has filmed cilia from ctenophores, polychaetes and bivalves as well as protozoa and has devised a means of expressing graphically the specific differences. Fig. 14A shows the beat of a compound cilium, or membranelle, of *Stentor*. During the effective stroke (ES) the ciliary shaft is held rigid and bending occurs only at the base.

When the ES is completed a return, or recovery stroke (RS) starts as a bending at the base which travels up the shaft until the cilium is again positioned at the beginning of the ES. The beat cycle of this particular organelle is complicated by the fact that the ES starts before the RS is completed. In *Paramecium* the beat cycle is slightly different in that there is no over-lap between recovery and effective strokes and the duration of the ES is much shorter than those of the RS. Sleigh (1968) has attempted a graphical representation of two beat characteristics. Fig. 14A shows this for the beat of a *Stentor* membranelle. The dotted line shows the angle between the membranelle and the body surface (a measure of the state of the ES) while the solid line shows the progression up the organelle of the RS bend. Thus at 25 msec from the start of these observations the basal inclination is 60° and the RS bend some 13 μm from the base. 15 msec later the basal inclination is 110° and the RS bend not yet started up the shaft. In contrast to this the same characteristics for the frontal cirrus of the hypotrich *Stylonichia* are shown in Fig. 14B. Although the beat does not appear very different at first glance, the graphical representation shows some interesting features, especially the pause in the middle of the ES which appears as a plateau in the curve for basal inclination. Note also the very rapid progression of the RS bend. These organelles are not in continuous movement and when at rest they are held with a basal inclination equal to that indicated by the plateau. Characterizing ciliary activity in this way has enabled us to compare and classify beat cycles of cilia from a wide variety of sources. We are also able to appreciate the fact that cilia do not all beat in the same way; there are many specializations each of which must have their functional significance. Our task now is to analyse these differences and to interpret them in terms of the animal's way of life.

Sleigh's graphs take into account only two dimensions. However, it has long been known that ciliary beat is not completely planar. Párducz (1966 for review) was able to 'freeze' the movement of cilia of a variety of protozoa by rapid fixation. These preparations suggest that in *Paramecium,* for example, the plane of the ES is different from that of the RS. In plan view the tip of the cilium describes a straight line in ES but returns through RS by an almost semicircular path. A more extreme view has been taken by Kuznicki *et al.* (1970) who, on evidence from ciné films, claims that in swimming *Paramecium* the ciliary beat is a travelling helical wave of short length much like that found in some flagella (e.g. *Euglena*). This controversial suggestion awaits independent confirmation. Indeed the whole field is ripe for detailed investigation, the concept of a typical ciliary beat being as dead as that of a typical flagellar beat.

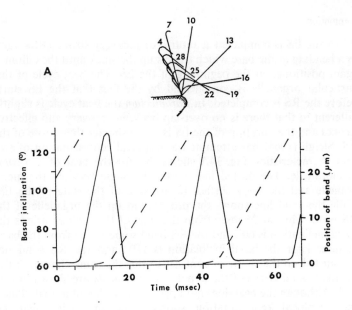

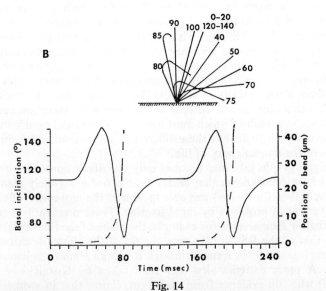

Fig. 14

Analysis of ciliary beat. **A**, membranelle of *Stentor polymorphus*; **B**, frontal cirrus of *Stylonychia mytilus*. In each case the upper figure shows the 'cilium' at various stages in its beat cycle. The figures indicate the time, in msec, during the cycle that the 'cilium' is in that position. The graphs show the basal inclination of the 'cilium' (unbroken line) and the distance of the recovery bend from the base (broken line) during the beat cycle. For further explanation see text. After Sleigh (1968).

COORDINATION BETWEEN CILIA

Flagella are usually borne either singly or in small numbers (except by some specialized flagellates such as the hypermastigotes) whereas cilia are often borne in fields of many hundreds covering the entire animal. The movement of a ciliate results from the sum of the contributions of the individual cilia. Such fields of cilia will exert their maximum propulsive power only if their ESs are aligned in the same direction. We know very little about the directional coordination of ciliary beat. In some cilia that invariably beat in one direction (e.g. some lamellibranch gill cilia and those of vertebrate ciliated epithelia) the central pairs of tubules are always aligned in the same direction. However, the cilia of ciliates usually have the ability to beat in various directions and in any case as we have seen the normal beat cycle may be three-dimensional.

There is also the question of coordination of the timing of the beat of individual cilia. Firstly, all the cilia on an animal tend to beat with the same frequency. Secondly, they beat in a coordinated way. There are three possible ways in which cilia could beat. They could beat randomly with no reference to neighbouring cilia, or they could beat altogether (synchronously). However, they more usually beat with metachronal rhythm, i.e. they all beat at the same rate but each cilium in a line is either slightly further or less far through its beat cycle than its neighbours (Fig. 15). Thus at any one time some cilia will be in their ES while others will be in RS but the interference that would result from random beating is reduced to a minimum. The advantage of this type of coordination over synchronous beating is that metachrony prevents the stop–go movement that would result if all the cilia beat together. A rowing 'eight' moves by synchronous beating of its oars, but its boat is large and the momentum carries it forward between strokes. A ciliate on the other hand is small and has almost no momentum and consequently would come to a halt between strokes. Metachronal beating ensures that the organism moves smoothly. It has also been suggested that synchronous beating would be difficult to coordinate whereas the movement of some triggering stimulus over the ciliary field makes metachronal coordination a relatively simple matter.

If one imagines a cilium (a in Fig. 15) with an effective stroke from left to right the cilium to the left (b in Fig. 15) may be either further (Fig. 15A) or less far (Fig. 15B) through its beat cycle. Thus there are at least two kinds of metachrony. Notice that in the former (Fig. 15A) the wave crest will move in the same direction as the metachronal wave crest. This is termed symplectic metachrony.

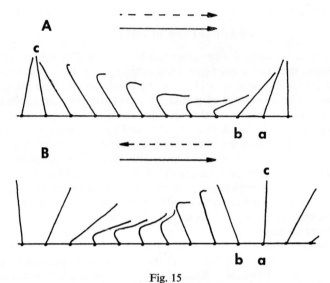

Fig. 15

Diagram showing cilia beating in metachronal rhythm. **A,** symplectic metachrony; **B,** antiplectic metachrony. Solid arrow, effective stroke; dashed arrow, direction in which the metachronal wave crest moves. For further explanation see text.

In Fig. 15B the metachronal wave crest will move in the opposite direction to the effective stroke; antiplectic metachrony. In both these types of rhythm it is assumed that the cilia are arranged in lines down which the wave crest passes and that in parallel lines the same thing is happening. Thus a cilium will be out of phase with its neighbours in its own line but in phase with those of lines alongside it. Two other sorts of metachrony occur in which the metachronal wave crest moves at right angles to the direction of the effective stroke (dexioplectic or laeoplectic). The direction of the ES and the metachronal wave crest may vary according to the physiological state of the animal but they do not vary independently. Thus the direction of the effective stroke can be inferred from the direction in which the wave crest moves.

It can be seen from Fig. 15A that in symplectic metachrony bunching and interference occurs during the ES. This type of rhythm is most often found in the cilia of epithelia involved in the movement of particles (e.g. frog gullet), and in organisms living parasitically in particle-laden environments (e.g. *Opalina* and *Isotricha*). Maybe tips of the cilia in ES combine their efforts to produce an efficient particle-moving system. Most free-swimming ciliates such as *Para-*

mecium show antiplectic metachrony, the cilia during ES being allowed an unhindered sweep through the water. Dexioplectic metachrony is found in the membranelles of *Stentor*.

CONTROL OF METACHRONY

The metachronal beating of cilia raises the problem as to how the cilia are coordinated. Since the end of the last century, controversy has raged back and forth with most of the protagonists favouring one or other of two main views. One theory suggests nerve-like or neuroid impulses passing along the ciliary rows. Such impulses are usually thought to travel along the membrane or along subpellicular fibre systems. The other theory maintains that the coordination is brought about mechanically, the movement of one cilium producing pressure changes in the medium that affect its neighbours.

The neuroid hypothesis rests largely upon the discovery of the stainable fibre systems that appear to connect ciliary bases. Such fibre systems include the kinetodesmata. It has been suggested that excitatory impulses could pass along these fibre systems stimulating the cilia one after another via their basal bodies. This idea was supported to some extent by experiments which showed that microincisions across ciliary rows resulted in a lack of coordination on the two sides of the cut. Such experiments are of course rather crude and may have many side effects. Párducz (1966) has shown that the kinetodesmata are very unlikely to be the neuroid organelles as their arrangement would not permit the complex ciliary behaviour that he has observed. For example in *Paramecium* the metachronal wave crests form a spiral round the animal and pass forward. At the posterior end of the animal the wave crest is almost at right angles to the kinetics (and hence the kinetodesmata) while in the oral groove the ciliary rows change direction such that they are now parallel to the metachronal wave crest. In *Nyctotherus* also the kinetal arrangement bears no relation to the pattern of metachronal wave crests. During ciliary reversal, the wave crest pattern may be even more complex. Thus it seems unlikely that a cut through the pellicle is disrupting coordination by destroying the continuity of the kinetodesmata, and one must look for other interpretations.

However, one series of experiments involving microsurgery must be considered more seriously. Taylor (1920) cut the fibres that extend from the anterior end of the AZM to the bases of the anal cirri of *Euplotes*. Usually these cirri beat in an intermittent but coordinated fashion. Taylor claimed that after cutting these fibres, which he termed 'neuromotor', this coordination was lost. Despite the specialized nature of these ciliary structures, the results of these experiments have been one of the major pieces of evidence in support of the neuroid hypothesis. Recently these experiments have been confirmed by Gliddon (1965) who reports that operation with both microneedle (Taylor's method) and radio-cautery results in uncoordinated crawling and avoiding reactions. However, Okijima and Kinosita (1966), in a series of very careful experiments, were able to show that after operation there was a period of 1–5 minutes during which coordination was poor, but that after that coordination and movement became normal and remained so. These workers attributed Taylor's results to the use of animals in poor physiological condition and suggest that the relatively high sodium concentration in the experimental medium aided recovery in their operated animals. Furthermore, Naitoh and Eckert (1969b) were able to show that the 'forward' or 'reversed' attitude of the cirri of *Euplotes* could be correlated with a hyperpolarization or a depolarization of the cell membrane, and that there is correlation remaining even after transection of the 'neuromotor' fibrils.

These fibres have been shown by electron microscopy to be composed of microtubules about 20 nm in diameter and Pitelka (1969) has pointed out that it would be odd for the tubules of *Euplotes* to have conducting properties while those of other ciliates do not. She also notes that many hypotrichs have similar fibres running from the end of AZM, but only in *Euplotes* do they appear to connect with the anal cirri. Thus in *Euplotes* only is there the physical possibility of coordination being effected via these fibres. It is much more likely that their function is one of support, as would seem to be the case for many of the elements of the infraciliature.

Although the majority of workers discount a coordinating role for the infraciliature, there are facts which are difficult to explain solely in terms of mechanical interaction between cilia. There must, for example, be some other mechanism controlling ciliary reversal and changes of swimming direction, and many of the complex movements of hypotrichs. Some of these special situations will be dealt with below. Nevertheless the weight of evidence concerning the propagation of metachronal waves seems to support a theory of mechanical interaction although there are some interesting exceptions.

Two very different sets of observations give us our most dramatic evidence in support of the mechanical theory. One comes from work with the ctenophore *Bereoe* (Sleigh and Jarman, 1972). This primitive metazoan has meridional rows of compound ciliary organelles called comb plates. Sleigh and Jarman made a small paddle-like device that could be placed between two individual comb plates. This paddle was then vibrated at a given frequency and films showed that its movement induced a slightly out of phase movement in the adjacent comb plate, in turn passed on to *its* neighbour and so along the line. They discovered that they could induce metachrony with the comb plates beating at various frequencies dictated by the rate of paddle vibration. This induction did not depend upon direct contact between the paddle and comb plate. The impression is that the movement of the paddle at first resulted in the comb plate being pulled passively by the drag of the medium. This passive movement appeared to trigger an active ES followed by an RS. The comb plate then remained inactive until the paddle induced first passive then active movement once more.

A second interesting piece of evidence comes from a flagellate *Mixotricha*. This creature lives in the guts of termites and was thought to possess a large number of flagella. However, Cleveland and Grimstone (1964) showed by electron microscopy that the animal has in fact only four flagella and that most of the surface is covered with a large number of spirochaete-like bacteria attached but not fused to the surface of the organism. These bacteria beat metachronally and propel the flagellate through its environment. In this case it is difficult to envisage any effective neuroid transmission in the pellicle. It is also interesting that Child (1965) was able to obtain good metachrony in lamellibranch gill cilia after they had been glycerol extracted and reactivated with ATP.

Further support for a mechanical linkage hypothesis comes from studying the effects of increased viscosity on the frequency of ciliary beat and the velocity of metachronal waves (Sleigh, 1966). If the viscosity of the medium is increased, e.g. with methyl cellulose, the frequency of ciliary beat decreases. If the progress of a metachronal wave depends upon the mechanical interaction of cilia, one would expect the reduced beat frequency would lead to reduced metachronal wave velocity. If, on the other hand, the metachronal wave results from some sort of stimulus propagated in or under the animal's surface, increase in viscosity should not affect wave velocity. In *Opalina** the former proposition holds (Fig. 16A). The same is true for *Paramecium,* but in this case there is a change from antiplectic to symplectic metachrony with increasing viscosity, a change more easily explained in terms of a changing mode of mechanical interaction than in terms of a neuroid hypothesis. However, for *Stentor* membranelles the evidence has other implications (Fig. 16B). In this case a slight increase in viscosity reduces the frequency of beat but

---

*Opalina is a symbiotic protozoan of the frog rectum. For many years it was considered to be a primitive ciliate but is now placed in a superclass of its own in the subphylum Sarcomastigophora. It has been extensively used in investigations of ciliary movement.

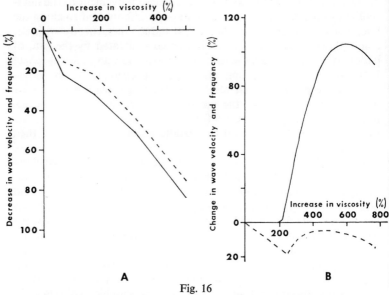

Fig. 16

Effect of increase in medium viscosity on the frequency of beat (broken line) and the velocity of the metachronal wave (unbroken line). **A,** *Opalina*; **B,** membranelles of *Stentor*. After Sleigh (1966).

has no effect on the wave velocity, suggesting that metachrony is controlled by a mechanism unaffected by viscosity. At higher viscosities, however, this type of control is replaced by a mechanical one, wave velocity falling off with increase in viscosity. The change is presumed to take place when the viscosity is such that the drag from one membranelle can affect the movement of its neighbour. The fact that the beat is dexioplectic may be the reason for the absense of sufficient drag at low viscosities. Other evidence suggests that the control of metachrony in *Stentor* may be neuroid. For example increased beat frequencies due to increase in Mg^{++} and Al^{+++} concentrations do not result in increased wave velocity. Also a cut across the AZM may result in reduced beat frequencies in the distal portion but wave velocity is again unchanged. If the membranelles are close together the wave velocity in mm/second is slower than when the membranelles are wider apart. This suggests an 'impulse' taking a constant time to go from one membranelle to the next.

Mainly on the basis of these observations of *Stentor*, Sleigh (1957) suggested that neither neuroid nor mechanical factors need be the sole controlling influences. He suggests that the viscous drag may trigger an internal excitation within a neighbouring cilium. Párducz (1966) on the other hand favours the hypothesis of 'waves of excitation of endogenous origin passing at regular intervals continuously along the periphery of the cell' and 'that the site of the metachronal impulse transmission is the cortical ectoplasm' although he stresses that no preformed fibrillar structures play a role in the process. Párducz took this attitude because he was not satisfied that mechanical effects could explain the complicated dynamic metachronal rhythms that his rapid fixation and staining methods had demonstrated.

It has been suggested that a wave of electrical excitation passing over the pellicle may explain metachronal coordination. However, this is not likely to be true for *Paramecium* as Eckert and Naitoh (1970) have shown that action potentials and passively conducted pulses in this animal spread at a rate of over 100 mm/sec as compared with a rate of less than 1 mm/sec for metachronal waves.

The present position is not wholly clear. However, it does seem safe to rule out any coordinating role for sub-pellicular fibril systems. It also seems likely that propagated action potentials play no part in normal metachrony. It is safe to say that in many ciliary systems mechanical interaction is the coordinating influence in metachrony. However, this may be mediated through intracellular excitation of the cilium and in some cases neuroid transmission may exist. Párducz's (1966) statement in favour of neuroid transmission in such animals as *Paramecium* seems not to discriminate between the coordination of cilia in metachrony and the control of the direction of the waves and their part in the animal's complex behaviour. Surely it is reasonable to suppose that in such animals, while the control of total behaviour is largely internal, metachrony itself may be largely mediated through mechanical interaction.

We know little about the internal control of the more complex pieces of locomotory behaviour with the exception of ciliary reversal. This reversal of the direction of the effective stroke is a common feature in ciliate movement and of course results in the animal swimming backwards. In many ciliates the cilia usually beat forwards or backwards but in some the direction of the ES may be continuously variable between the two extremes. In *Opalina* the cilia beat towards the posterior in symplectic metachrony. When the animal turns or becomes 'excited' the plane of the effective stroke moves clockwise. This change is easily visible as a change in the direction of movement of the metachronal wave crests. The deviation from normal may be from a few to 180 degrees. Kinosita (1954) has inserted microelectrodes into *Opalina* and investigated the relationship between the direction of the effective stroke (as manifested by the direction of the metachronal wave crests) and the membrane potential. He found an

almost exactly proportional relationship between the membrane
potential and the direction of effective stroke (Fig. 17). He also

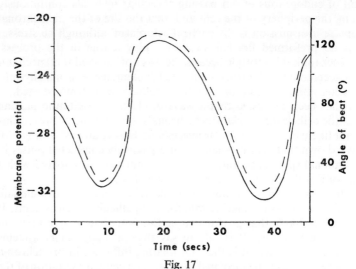

Fig. 17

Relation between direction of metachronal wave movement (unbroken
line) and membrane potential (broken line) in *Opalina*. After Kinosita
(1954).

studied the effects of substituting KCl for the Ringer's solution that
normally bathed the experimental animals. The results of such a
change were (1) at first the polarity of the membrane potential was
reversed to become positive inside, though this fell slowly to zero
potential, and (2) the ciliary beat reversed initially but gradually
returned to normal. After return to normal Ringer's the potential re-
established itself at its former negative value. If the animals were
first subjected to calcium-free Ringer's solution and then transferred
to KCl, the latter solution produced an abolition of the membrane
potential with only a very slight change in the angle of effective
stroke. If the treatment was repeated the second immersion in KCl
produced membrane depolarization but no change in ciliary beat.
These experiments established that in *Opalina*, at least, the angle of
ciliary beat is intimately associated with the membrane potential but
that changes in the potential are not the cause of the changes in beat.
This is especially true in animals from which calcium has been leach-
ed. It could be that changes in membrane potential normally result
in some movement of calcium from one cellular compartment to
another which in turn results in a change in ciliary beat. Depletion
of cellular calcium might disrupt this chain of events. On the other

hand the changes in membrane potential may play no active part in the process, merely reflecting the events bringing about the changes in beat.

In *Paramecium* Naitoh and Eckert (1969a) have rather more direct evidence implicating electrical events in ciliary reversal. They inserted a microelectrode into a *Paramecium* and stimulated it mechanically at either the anterior or posterior end. In the former case this resulted in a transient increase in membrane permeability to calcium. This calcium flow produces a temporary depolarization which elicits ciliary reversal. Stimulation at the posterior end results in hyper-polarization. These results explain in terms of ion fluxes and excitable membranes the behaviour observed in the avoiding reaction. They also suggest that depolarization is more likely to be a cause than an effect of ciliary reversal.

It has been known for many years that calcium plays an important part in ciliary reversal. Kamada (1940), from work with *Paramecium*, suggested that ciliary reversal resulted when an unknown anion X was released from a compound CaX. This breakdown is supposed to result when the intracellular calcium level drops. Kinosita believes that his results from *Opalina* support this hypothesis and that when the animal is placed in KCl the positive overshoot is due to a rapid efflux of calcium. However, if Kamada's suggestion is correct then either X quite quickly disappears as a free anion, or its action is only temporary, as ciliary reversal in KCl is transitory and cannot be illicited at all in animals with a low intracellular calcium concentration. An alternative and more attractive hypothesis has been put forward by Jahn (1962) on the basis of his reinterpretation of the results of Kamada and Kinosita (1940). The Japanese workers had accumulated a large quantity of data on the duration of ciliary reversal in solutions containing various concentrations and ratios of calcium and potassium. They found that no one value of the ratio [K]/[Ca] produced maximum duration of reversal. Jahn in his examination of these results showed that maximum duration of reversal occurred in solutions with a constant value for $[K]/\sqrt{[Ca]}$, the Gibbs–Donnan ratio. Because of the different properties of mono- and divalent ions, solutions with the same Gibbs–Donnan ratio result in a constant value of [K]/[Ca] for ions bound to fixed anionic sites such as those on proteins. This property is independent of the concentration of the solutions. This has a number of consequences. Firstly to maintain a constant ratio for [K]/[Ca] at the binding sites one must maintain a constant value for $[K]/\sqrt{[Ca]}$ in the surrounding solution. Secondly dilution of the medium will result in a change in its Gibbs–Donnan ratio and hence in the ratio of the ions bound to anionic sites. Thirdly the addition of a given amount of calcium to

the medium will affect the Gibbs–Donnan ratio more at low concentration than at high. Jahn suggested that maximum duration of
ciliary reversal resulted at a certain ratio of bound calcium and
potassium and at some corresponding Gibbs–Donnan ratio for the
surrounding solution. The site sensitive to the ions may be the outer
membrane or perhaps some structure in the cilium itself.

Jahn's idea that some optimum value for bound calcium rather
than a simple diminution of bound calcium results in maximum
ciliary reversal is supported by the work of Grebecki (1965). He
treated *Paramecium* with various concentrations of EDTA, a substance that chelates calcium. He found that at low concentrations of
EDTA the animals behaved normally. However, if he increased the
concentration they swam alternately forwards and backwards, a behaviour termed periodic ciliary reversal. Further increase in EDTA
concentration results in circular swimming movement in which some
of the cilia beat forwards and some backwards. Still further increase
results in the re-establishment of normal forward swimming. The
greater the concentration of calcium in the medium before the experiment, the greater is the concentration of EDTA required to induce
the onset of any of these behavioural responses. If *Paramecium* are
decalcified so strongly that they have returned to normal forward
swimming, reversal can be induced by *adding* calcium. In fact, one
can reverse the series of responses seen during decalcification by
gradually adding more and more calcium to the medium of these
EDTA-treated animals. These and other experiments performed by
Grebecki suggest that ciliary reversal occurs at some particular
value for intracellular calcium concentration and that values above
and below this result in partial reversal grading into normal movement. It is also apparent from these experiments that ciliary reversal
is not an all-or-nothing reaction on the part of the animal. Intermediate states exist where reversal is partial either in time (periodic
ciliary reversal) or over the animal's surface (circular swimming).

More recently Naitoh (1968) has proposed that ciliary reversal is
essentially an excited state stimulated by calcium. His results suggest,
as do those of other workers, that reversal results when calcium is
displaced from anionic binding sites. However, he differs from previous workers in suggesting that the calcium once displaced moves
to some part of the ciliary apparatus where it initiates reversal. This
suggestion is supported by the fact that glycerinated models* of

*If a ciliate is placed in a 50 per cent solution of glycerol at a temperature
below 0°C most of the water soluble constituents of the animal are leeched out
leaving a largely protein 'skeleton'. This preparation is often termed a 'glycerinated model' and will still have the gross morphology of the animal and will still
show some contractile activity in the presence of suitable substances, primarily
ATP and calcium.

ciliates show reversal in the presence of ATP and calcium. However, this hypothesis does not explain why injections of calcium-binding oxalate produce ciliary reversal, for although this would result in calcium coming off binding sites, it would precipitate as calcium oxalate and not be available to stimulate ciliary sites. Nor does it explain why externally added calcium does not normally result in ciliary reversal. The answers to many of the fascinating problems of ciliary reversal still evade us but further investigation into the role, distribution and movement of calcium will prove fruitful.

To regard ciliary reversal as an excited state is not a new idea, mainly because it can be induced by electrical stimulation. If *Paramecium* is placed in an electric field the cilia on the part of the animal directed towards the cathode reverse the direction of their beat while those at the anodal end beat in the normal way. The stronger the current the larger is the area of reversed cilia. It has been assumed from these observations that depolarization of the membrane results in excitation leading to ciliary reversal while hyperpolarization leads to augmentation of the normal beat. There have been a great many experiments conducted to examine this phenomenon (see Kinosita and Murikami, 1967, for a useful review) and several have attempted analogies with the classical concepts of nerve physiology (e.g. Jahn, 1961). Dryl (1970) has considered the problems of excitability in the protozoan cell. As he points out, some ciliates have retained both receptor and effector functions. Indeed the membrane is the site of reception of external stimuli and also can produce graded or all-or-none responses, as for example during various types of swimming. For instance Kinosita *et al.* (1964) found that spontaneous reversal of cilia resulted in a depolarization while contraction of the cell (*Paramecium*) coincided with a hyperpolarization. If the animal was made to perform periodic ciliary reversal by immersion in a mixture of barium and calcium chloride solutions, each reversal was accompanied by a short period of strong depolarization identical in appearance to the action potentials found in other types of excitable cell. It is assumed that these spikes are propagated.

STRUCTURAL ANATOMY AND BIOCHEMISTRY OF CILIA

While many workers have concerned themselves with the problems of coordination of cilia, others have been probing the mechanisms controlling their cyclical contractile activity. Much of the effort has gone into trying to evaluate the roles of the various ultrastructural elements. One of the earliest questions asked was: which are the contractile elements, the peripheral fibrils or the central pair? (See p. 46 for details of ciliary fine structure.) The central pair seem not well placed for such a role and it is more likely that they are skeletal. However, Harris (1961) estimated that if they alone were to act in maintaining the cilium's rigidity, they would need a Young's modulus close to that of steel wire, making them the strongest of biological materials. It is much more likely that the stiffness is supplied not only by these fibrils but also by the turgor within the cilium. Harris calculated that, assuming a reasonable Young's modulus for the membrane, a concentration difference of 30 mM between inside and out would produce sufficient turgor; a not unreasonable state of affairs. Shrinkage of cilia at fixation and their sensitivity to change in external osmotic pressure also argue for rigidity through turgor. It has also been suggested that the central pair play a part in conduction of a contraction stimulus up the cilium. However, there is no evidence for this and many sensory cilia (well able to conduct information) have no central pair. Sleigh (1961) suggests that these fibrils, which are attached by radial connections to the peripheral doublets, ensure that contraction in the latter results in a properly oriented bending rather than a general distortion or spiralling.

It is now generally accepted that it is the outer doublets that produce the bending and that their contractile mechanism probably involves a relative sliding rather than an absolute shortening (see Chapter 2, p. 49). Sleigh (1968) has calculated that each peripheral doublet must slide about 10 nm relative to its neighbour for each 10° of bend. Thus during an average ES adjacent doublets will slide about 80 nm over one another. He also calculated that the rate of slippage would be from 1–10 μm/sec, a rate comparable with that of the filaments of striated muscle.

Recently it has become possible to obtain some idea of the types of protein that make up the ciliary substructure (Child, 1967). For example Gibbons (1965) showed that the arms of peripheral doublets contain ATPase. Gibbons and Rowe (1965) showed that this ATPase, which they called dynein, is a polymer composed of globular subunits about 9 nm in diameter and about 14 nm long. When extracted, this protein is usually in the form of a short chain of subunits, the spacing of which is similar to the periodic structures seen in longitudinal sections (Gibbons and Grimstone, 1960; Hopkins, 1970). It is thought that two such rows of these subunits arranged longitudinally on the A subfibril make up the arms. In its properties and structure dynein is similar to myxomyosin, a protein associated with protoplasmic streaming in slime moulds. It is also similar to the proteins of the mitotic spindles and Gibbons and Rowe think that it may be one of a class of proteins that play an important part in cell motility but which are separate from contractile proteins related to myosin.

By similar techniques Renaud et al. (1968) have shown that the protein making up the doublets, except for the arms, is similar to actin in a number of ways including its aminoacid composition. This protein has been named tubulin and has a molecular weight of about 60 000. Stevens (1970) contests the homology with actin, mainly on the grounds that the proteins contain many dissimilar sequences of aminoacids. Renaud *et al.* (1968) also showed that the sub-units were arranged linearly to form 4 nm diameter fibrils. Grimstone and Klug (1966) examined the fine structure of doublets from *Trychonympha* and showed that each individual tubule was composed of some 12 fibrils which had a beaded appearance. The spacing of beads was 4 nm from centre to centre while the space between adjacent fibrils is 4–5 nm. The beads of adjacent fibrils do not lie directly alongside one another, but are slightly staggered so as to produce a helical surface pattern (see Fig. 18). This pattern is further compli-

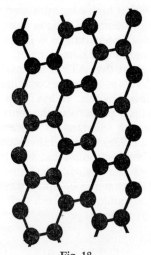

Fig. 18
Suggested arrangement of subunits in the wall of a microtubule. After Grimstone and Klug (1966).

cated by the fact that the fibrils are not exactly linear but slightly zig-zag.

Although the details are still in doubt it seems reasonable to suggest that the molecular architecture of the ciliary doublet is as follows. It is composed of 2 tubular subunits one of which is complete while the other attached to it is C-shaped in section. The complete one is composed of about 12 filaments which are made up of a linear arrangement of tubulin molecules. These are globular and about 3·5 nm in diameter. Two rows of dynein molecules perhaps joined end to end and attached every 12 nm down the length of the A tubule (Hopkins, 1970). It seems likely that the dynein interacts with the tubulin proteins of the B microtubule of the adjacent doublet. This results in a relative movement of the doublets which produces bending of the cilium. This activity is probably powered by ATP hydrolyzed by dynein. Gibbons (1966) has noted the interesting fact that one molecule of dynein splits 13–35 ATP molecules per second which is approximately the same as the frequency of ciliary beat. Perhaps each dynein molecule splits only one ATP molecule in each beat cycle.

ATP has been shown to stimulate beating in cilia and flagella. Brokaw (1961) found that the beat of isolated *Polytoma* flagella, induced by ATP, is very similar to that of flagella on the intact organism. The frequency of beat is more or less dependent on the ATP concentration, but at high concentrations the amplitude of the beat is so small that movement is difficult to detect. The rate of dephosphorylation of such preparations is in the same range as that reported for myosin and muscle homogenates.

Although it is apparent that ATP is necessary for normal movement, the reason for ATP's induction of coordinated wave production in isolated cilia is less obvious. It has been suggested that this rhythmical movement is a property of the contractile material. Thus Pautard (1962) observed spontaneous vibration in flagellar gel. Kinosita and Kamada (1939) proposed that stretching of one side of the cilium stimulates contraction on that side, and this idea is also a prerequisite of one of the mathematical models of ciliary beat. However, it is unlikely that the complex wave forms reported by Sleigh (1968) (see p. 66) could result from such a simple mechanism, and the control of ciliary beat is probably much more subtle.

CONTRACTION

Many ciliates are capable of actively changing their body shape by contraction. *Lacrymaria* (Fig. 1C) can wave its proboscis and *Stentor* can bend to and fro about its point of attachment. These movements are relatively slow and to date we know little about them. Some ciliates, however, are capable of much faster and violent movements. The best known examples are the heterotrichs *Stentor* and *Spirostomum* and the peritrichs, in which the zooid and sometimes the stalk are contractile.

In *Spirostomum* contraction results in the normally elongated cylindrical animal becoming ovoidal. The contraction is not localized to any particular part of the animal and is simply a very rapid shortening. This movement can be readily demonstrated by tapping a dish or slide containing the animals. It can also be elicited by a wide variety of chemical, thermal, electrical and mechanical stimuli. Indeed so sensitive is the contractile system that it is extremely difficult to fix these organisms in a relaxed state. Boggs (1965) has fixed them in an extended position by allowing them to swim in a gelatine solution which gradually hardened holding the animals immobile. The gelatine blocks with their trapped animals were then dropped into fixative. Legrand (1968) claims that animals first treated with the anaesthetic chloretone can be fixed with only slight contraction. Anaesthetics may well change structure, however, and investigators should be aware of possible artifacts. For instance many anaesthetics disrupt microtubular elements (Allison *et al.,* 1970) which in heterotrichs are probably involved in movement (Bannister and Tatchill, 1968).

Contractions can be very quick. High-speed filming has shown that *Spirostomum* contracts to about 50 per cent of its original length in 4 msec (Jones *et al.,* 1966). Strong stimulation can result in further shortening but when the animal is about one third of its original length it often bursts.

In *Stentor,* like *Spirostomum,* contraction involves the whole of the animal, but the contractile ability is much greater in the posterior region. When extended it is trumpet-shaped but when contracted it is spherical. If the animal is attached, contraction results in the animal being pulled down to the substrate. *Stentor* has at least two contractile systems. One is made up of the longitudinal myonemes that produce the shortening of the animal while the other runs round the animal's anterior under the AZM and on contraction draws in the oral ciliature. These two systems have a certain amount of independence. As in *Spirostomum* the rate of contraction is high, and a shortening to 20 per cent or less of original length can take place in 10 msec (Jones *et al.,* 1970a).

Peritrichs have two contractile systems. One is found in the zooid. Here there are myonemes very like those of heterotrichs (see p. 54) running both longitudinally and circumferentially. These result, upon contraction, in the zooid becoming spherical. The region bearing the oral cilia is completely inverted and protected by contraction of a myoneme which acts like a purse-string round the anterior of the zooid. Some genera possess a second contractile system in the stalk. The best known examples are the solitary *Vorticella* and the colonial *Carchesium* and *Zoothamnium* (Plate 2A). In these animals the stalk myoneme runs down inside the stalk, enclosed by a membrane together with mitochondria that supply the energy for contraction. The stalk myoneme passes upwards into the zooid with whose myonemes it is continuous. Despite this continuity there appears to be some degree of independence in the two systems. The regions of the stalk outside the myoneme contain fibrous or tubular elements which presumably act as antagonists to the contraction. Similar supporting material is found in the non-contractile stalks but in these cases the myoneme is lacking. Because the outer part of the stalk cannot shorten, contraction of the myoneme results in its distortion into the characteristic spiral form. The movement from the extended to the contracted state looks just like a stretched spring returning to a tight coil. As this happens the zooid is pulled down towards the substrate, and the effective length of the stalk may shorten to less than 30 per cent of its original length. This movement is completed in less than 10 msec (Jones *et al.*, 1970b).

We know relatively little about the chemistry of the myonemes. They are proteinaceous but whether or not they contain actinomyosin-like substances has not been discovered. In some ways, however, they resemble contractile materials of higher animals. For instance it has been firmly established that calcium is involved in the initiation of contraction as in muscle. *Spirostomum* contracts less fast in calcium-free solutions and more quickly in calcium-rich solutions (Jones *et al.*, 1966). More recently Sleigh (1970) has examined the effect of various chemicals on the excitation threshold and the frequency of spontaneous contraction. He concluded that excitability depended on the level of free calcium in the cytoplasm. If this level dropped the animals became less sensitive to a standard stimulus or less likely to contract spontaneously. For example, less than 10 per cent of animals respond to a constant electrical stimulus in a solution of 5×10^{-6} M calcium chloride whereas about 70 per cent responded to the same stimulus in a 5×10^{-5} M solution. Similarly raising magnesium and barium concentrations results in greater excitability. It is suggested that these ions may replace calcium in some of its functions, making it available for the initiation of contraction. Caffeine and nicotine, which are thought to release calcium from binding sites and to inhibit uptake of calcium by the sarcoplasmic reticulum of striated muscle, increase the excitability of *Spirostomum* and it is presumed that their action here is similar to that in muscle. Ettienne (1970) injected aequorin into *Spirostomum*. This substance, extracted from jellyfish, will emit light only in the presense of calcium. During electrically stimulated contraction Ettienne demonstrated that light was emitted from this animal suggesting a rise in the level of free calcium in the cytoplasm at that time.

In vertebrate striated muscle relaxation is brought about by a resorption of calcium into the membrane-bound vesicles of the sarcoplasmic reticulum. There is some evidence of a similar calcium binding system in ciliate myonemes. Carasso and Favard (1966) have detected calcium deposited in canals inside the myoneme of peritrich stalks. Heterotrichs also have vesicles intimately associated with the contractile material and Vivier *et al.* (1969) have suggested that in *Spirostomum* these vesicles accumulate calcium in their relaxing role and that it is then precipitated to form apatite crystals (see p. 60). Ettienne (1970) has shown the presence of calcium in the membrane-bound vesicles by precipitation with oxalate. He has further shown that the calcium which initiates contraction is unlikely to enter through the pellicle but is much more likely to be released from such vesicles, being resorbed during relaxation.

As yet we know very little about the events which come between the stimulus and response. Sugi (1960), working with *Carchesium,* has shown that if two microelectrodes are placed close to the stalk, an impulse from these can stimulate stalk contraction. However, the contraction always starts at the stalk/zooid junction. Thus it appears that the zooid has control over stalk contraction; indeed stalks without zooids do not respond to stimulation. The evidence does indicate, however, that the stalk is sensitive to electrical stimulation but that they relay the information to the zooid. The contraction once initiated at the stalk/zooid junction is propagated down the stalk at a constant rate. Other evidence that a stimulus for contraction may pass along the myoneme comes from a comparison of the behaviour and structure of *Carchesium* and *Zoothamnium.* In the former the stalk myonemes of the members of the colony are not connected. Members of such a colony are capable of independent contraction. In *Zoothamnium* the stalk myoneme is continuous and when contraction occurs it involves the whole colony. How this communication is effected is not known but it may be mechanical, the myoneme responding to contraction in an adjacent area.

Some work has been done with glycerinated models of *Vorticella.* Levine (1956) showed that calcium will produce contraction in such preparations and that EDTA would reverse the process. Hoffman-Berling (1958) was able to show that glycerinated stalks contract at calcium concentrations as low as 1×10^{-6} M. He showed that ATP would relax stalks previously contracted by the addition of calcium. This latter experiment may show ATP acting as a chelating agent, binding calcium, as it has been shown to do in muscle preparations. (For a review of work on glycerinated muscle see Bendall, 1968.) Townes and Brown (1965) showed that raising the pH would induce contraction in glycerinated *Vorticella* provided that calcium, magnesium and ATP were present. They also showed that EDTA would produce relaxation. Thus it appears that the role of ATP has not yet been definitely established as contraction may occur in its absence.

Recently Amos (1971) has confirmed Hoffmann-Berling's observation that glycerinated *Vorticella* stalks will contract in the presence
of calcium and relax when calcium is removed, all in the absence of
ATP. He found that glycerinated stalks stay extended if there is less
than 10^{-7} g ions/l of calcium in the experimental medium. At higher
levels the stalks contracted and stayed contracted so long as the
calcium level was maintained above the threshold concentration. He
suggests that in life contraction is initiated by the release of calcium
from the myoneme canals and relaxation follows if the calcium is
taken back into the canals. The resorption of calcium is dependent
upon ATP. The energy for the contraction itself could come from
the chemical potential of the calcium ions. He calculates that the
amount of calcium needed to bring the myoneme from the relaxed
to the contracted state would be equivalent to about a $1 \cdot 1$ mM solution
of calcium chloride of the same volume as the myoneme. The
physical properties of the myoneme suggest that it may be composed
of a rubber-like substance, lengthening and shortening being brought
about by the formation and breaking of calcium sensitive cross-
links (Weis-Fogh and Amos, 1972).

EXTENSION

The extension of a *Vorticella* or *Stentor* after contraction involves
two processes. The first is the relaxation of the myonemes and the
second is the restoration of original shape which is brought about
by the antagonists to the myoneme. As we have already noted, the
relaxation of the myoneme may proceed in much the same way as in
muscle. In peritrichs the antagonists of the contractile element of the
stalk are probably the fibrils and tubules of the annulus, and perhaps the sheath and turgidity of the stalk. However, the rate of
extension would seem to be limited by the relaxation of the myoneme
which is apparently relatively slow in peritrichs (Jones and Morley,
1969). In heterotrichs, however, the relaxation of the myonemes is
rapid. In *Stentor* one can see the development of bends in them as
they slacken off after contraction and before extension occurs. The
rate of extension seems to be limited by the antagonists. One possible
explanation has been put forward by Bannister and Tatchell (1968)
who suggested that extension in *Stentor* may be controlled by cross
connections between the sheets of microtubules that make up the
km fibre (see p. 53 and Plate 2B). These sheets must |slide over one
another during extension and contraction, and connections between
the sheets may prevent or slow changes of shape. They may also be
a means of active extension if relative sliding of these microtubules
could occur as it does in cilia.

4

FEEDING AND NUTRITION

No ciliate is capable of photosynthesis and none are chemoauto-
trophs. Therefore they all require a supply of organic material to
sustain life. This they obtain in a variety of ways and in various forms.
Some feed on solid matter, alive or dead, and their nutrition is
termed holozoic. In others dissolved nutrients are absorbed through
the animal's surface; these are saprozoic. A few ciliates apparently
obtain food from photosynthetic algal cells living in their cytoplasm.
It is likely that many ciliates are both holozoic and saprozoic, and
many normally holozoic organisms can be persuaded to become
wholly saprozoic. Ciliate feeding can be further classified on the
basis of the type of food and the methods used to obtain it.

PARTICLE FEEDERS

Included in this category are those ciliates that feed on particles,
both dead and alive, that are small compared with themselves. These
particles may be organic detritus derived from dead animals and
plants or their wastes. In some habitats such material may be present
in considerable quantities (e.g. activated sludge, see p. 177). In these
habitats it is likely that both dissolved nutrients and bacteria will
also be present, and may be taken in as well as the detritus. Many
ciliates are able to live on bacteria alone. Cultures of *Paramecium,
Vorticella, Carchesium* and many others can be maintained by feeding
with bacteria such as *Aerobacter aerogenes* or *Bacillus subtilis.* Some
bacteria are apparently more nutritious, or perhaps less noxious,
than others as they vary in their ability to promote growth and reproduction. Pringsheim (1928) showed that *P. bursaria,* although able
to utilize *B. subtilis,* did not reproduce when supplied with *B.*

fluorescens or *Azobacter* sp. The latter may be indigestible because of its gelatinous coat. *B. proteus* supported growth for a while but not indefinitely. Hargitt and Fray (1917) found that *Paramecium* grew best when supplied with a mixture of bacteria rather than with a single species. However, this may be due to the limited range of bacteria tested as Philpot (1928) obtained best growth when only certain strains of *B. pyocyaneus* were given, indicating that a mixed diet was not beneficial under his conditions of culture. *B. pyocyaneus* is a pathogenic species and the fact that it is eaten by *Paramecium* illustrates the importance of ciliates in sewage treatment and possibly in our waterways. The numbers of bacteria eaten are large; Ludwig (1928) estimated that a *P. caudatum* requires about 1500 *B. subtilis* per hour for healthy growth, while Curds and Cockburn (1968) have shown that a *Tetrahymena* consumes about 600 bacteria an hour.

Particulate food may include micro-organisms other than bacteria. Yeasts, for instance, are readily eaten by many ciliates. Some will take both green and colourless flagellates, and many large ciliates will eat their smaller relatives; *Stentor coeruleus,* for example, can be grown on a diet of *Colpidum* (De Terra, 1966). *Bursaria* is perhaps the most voracious of the particle feeders, taking whole paramecia, some 200 μm or more in length. This type of feeding could perhaps better be termed carnivorous.

HERBIVORES

Some ciliates feed on unicellular algae both motile and non-motile. These animals use feeding methods similar to other particle feeders; the plants are swept into the cytopharynx by ciliary action and enclosed in a food vacuole. If the algae are large there may be but one to a food vacuole. Other ciliates feed on filamentous algae and their feeding processes are more specialized. *Frontonia leucas* is reported able to eat algae filaments up to 6 or 8 times its own length, coiling them up inside itself (Sandon, 1932). *Nassula* (Fig. 2A) feeds on filaments of blue–green algae (it can be cultured on *Phormidium inundatum*). Tucker (1968) has examined this animal and its method of feeding in some detail. This creature is a cyrtophorine gymnostome and has mouthparts composed of trichites, or rods, with no obvious oral cilia. The rods, from 23 to 34 depending on the individual,

---

Fig. 19

Reconstruction of the pharyngeal basket of *Nassula*. cr, crest; lcb, lower circumferential band; r, rod; sh, sheath; ucb, upper circumferential band. After Tucker (1968).

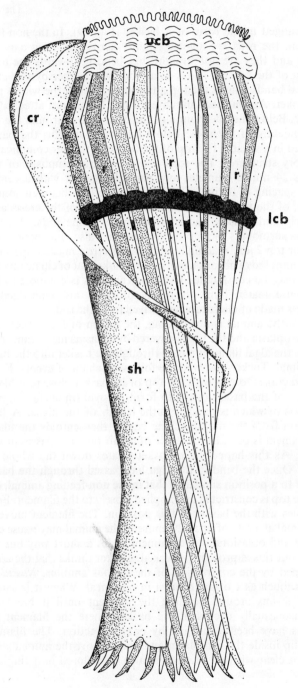

Fig. 19

are arranged in the form of a basket (Fig. 19). In the non-feeding position, the rods are aligned slightly across the long axis of the basket and they are not straight, but bend in towards the top and bottom of the basket. The rods are bound together by two circumferential bands, one at the top and one just below the widest part of the basket. Above these annuli each rod has a crest attached to its outside. Below the lower annulus the crests spiral around the basket independently of the rods. Below the lower annulus the basket is enclosed in a sheath. The top of the basket is further complicated by accessory structures. The rods and crests are composed of microtubules 24 nm in diameter and packed together with a centre-to-centre spacing of 36 nm. There are connections between some, but not all, of the microtubules. The microtubules of the crests are less regularly and densely packed than those of the rods. The upper annulus surmounts the top of the rods and also runs between them for their top 2 μm. Examination in the electron microscope shows it to be composed of fibrous material not unlike that of ciliate myonemes and it may be contractile. The lower annulus is composed of electron-dense material in which microtubules are embedded. The sheath is made of microtubules derived from the rods.

When the animal is not feeding, the lumen of the basket is filled with cytoplasm and the top is covered by a plasma membrane. During feeding the algal filament passes through the basket into the body of the animal. Tucker observed the following chain of events. First the *Nassula* comes to rest with the top of its basket close to a filament. The top of the basket opens to form an oval (in surface view) the long axis of which is parallel to the length of the algae. A hyaline extrusion from the mouth of the basket then enfolds the filament. The filament is bent, to the shape of a hair pin, and passes down the basket. As this happens the basket dilates down the whole of its length. Once the bend in the alga has passed through the basket it returns to a position similar to that in the non-feeding animal, except that the top is constricted and applied closely to the filament. Feeding continues with the basket in this position. The filament moves into the animal at a rate of up to 20 μm/sec. The animal may pause during feeding and occasionally the filament slides a short way out of the basket but this is promptly checked. Tucker thinks that the filament is gripped by the contraction of the fibrous annulus. *Nassula* may take as much as 1·6 mm of alga in one meal. When it is sated the animal swims around bending the filament until it breaks. The severance usually occurs in the basket where the filament might perhaps have been weakened by enzymic action. The filament is coiled up inside the cytoplasm and may distort the animal's shape. It is not clear whether or not the alga is enclosed in a single food vacuole or in a series. However, enzymic action quickly results in the break-up of the filament into short lengths. This process is completed in about 10 minutes and the separate pieces are enclosed in individual food vacuoles. Tucker observed many vesicles, like the primary lysosomes seen by the other workers (see p. 99), in the vicinity of the basket and these may contribute to the rapid disintegration of the filament.

How filaments are moved into *Nassula* is not clear. Suction is unlikely to be important and active movements of the basket cannot be detected except at the onset of feeding. The basket probably holds the filament during pauses in ingestion. The most likely explanation is that the filaments are pulled in by cytoplasmic movements. Cytoplasm could enter the basket between the rods at the top and sweep down inside pulling the filament.

CARNIVORES

Many ciliates, both large and small, are carnivores. In many cases the prey is small and is simply swept into the cytopharynx and enclosed in a food vacuole. However, some ciliates have large prey and then have elaborate mechanisms for capture and ingestion. The best known examples are rhabdophorine gymnostomes, e.g. *Didinium, Coleps* (Fig. 1A), or suctorians, e.g. *Podophrya* (Fig. 3C).

Didinium is a voracious killer of *Paramecium* (Plate 3). This animal's feeding has been the subject of much published work culminating in the recent work of Wessenberg and Antipa (1970), upon which the following account is based. Unfed *Didinium* swim rapidly through the water until they happen to collide with a *Paramecium*. They swim in a straight line and appear not to perceive the prey until the collision and there is no evidence of chemoattraction. When contact is made *Didinium* discharges two sorts of organelles into the prey. These are called toxicysts and pexicysts. The former are extruded up to 40 μm and are disposed in a ring around the proboscis. The pexicysts are inside this ring and are discharged only some 2 or 3 μm. The discharge of these organelles is so violent that the predator is pushed slightly away from the prey. This recoil tears the pexicysts from the *Didinium* which remain embedded in the *Paramecium*. The toxicysts, stabbed deep into the prey, probably contain poison. This results in the lysis of the cytoplasm and a cessation of ciliary movement close to the strike. The toxicysts are pulled into the *Didinium* thus pulling the *Paramecium* towards the proboscis which dilates to allow the prey to pass into the predator. This pulling first of the toxicysts and the attached prey and later the prey itself is apparently accomplished by cytoplasmic movement. Streaming is visible in the animal; towards the anterior around the periphery then running towards the posterior in the interior, pulling the toxicysts with it. Once the harpooned animal is inside the cytopharynx it, too, comes under the pulling influence of the streaming cytoplasm.

Once the *Paramecium* is inside the *Didinium* the cytopharynx closes and the proboscis reforms. The *Didinium* then swims to a convenient 'resting place' where it rotates gently until, after about two hours, its meal is digested. Didinia that have made a strike but lost their prey also 'rest' presumably to remake and position their extrusion organelles. Paramecia that have been struck but which escape invariably die later, presumably from the effects of the toxin.

Butzel and Bolten (1968) found that starved paramecia are an insufficient diet for *Didinium*. Paramecia fed with the bacteria *Aerobacter* or *Serratia* supported active growth of *Didinium*. The predator needs 2 or 3 prey before it will divide. With at least two hours between meals this means a maximum growth rate of about 3 divisions per day. These authors noted that didinia allowed to eat 3 prey always divided. Those fed only 2 prey either divided or encysted while those given a single prey did neither (as did those fed starved paramecia). Encystment in this ciliate appears to be a response to a feeding level just insufficient for division rather than in response to starvation or dietary insufficiency.

Dileptus anser is another predatory gymnostome. Its proboscis is also armed with toxic trichocysts (Dragesco, 1962). As it swims, it waves its mobile proboscis, increasing the chance of a collision with suitable prey. It is less conservative in its diet than *Didinium* and will take a variety of ciliates, flagellates, amoebae and even flatworms. When struck by the proboscis the prey is either killed or injured. Slightly harmed animals may escape by twisting round and round until they break away (Miller, 1968). Presumably escapees are able, in most cases, to regenerate the damaged parts. Damage to the prey takes the form of lysis of the struck part. The injured part sticks to the proboscis, and the attached prey is moved to the cytopharynx which dilates to allow it to be taken into the cell. It seems that, as with *Nassula* and *Didinium*, cytoplasmic streaming is responsible for pulling the food through the cytopharynx.

The suctorians are almost all carnivores. They are sessile animals that catch free-swimming ciliates on the end of tentacles which are then used like 'drinking straws' to transfer the contents of the prey to the predator. The following account of suctorian feeding draws on many original reports especially those of Kitching (1952a), Hull (1961a,b) and Rudzinska (1965, 1970) which are concerned with two species, *Podophrya collini* and *Tokophrya infusionum*.

Although suctoria will eat a wide range of ciliates, there is some selection of prey, or at least the suctorians are unable to catch various ciliates, especially large ones such as *Stentor* or those with thick pellicles such as *Euplotes*. Hull (1961a) showed that these animals did adhere to the tentacles for a short time but quickly broke away. The usual food seems to consist of medium-sized, soft-bodied forms like *Tetrahymena* and *Paramecium*. The prey swims on to the motionless tentacles of the suctorian and adheres to the tips. Its cilia are paralysed in the region of the strike but the other cilia continue to beat for some time, and their movement may bring the prey into contact with other tentacles to which it also adheres; it may be finally attached to as many as seven. The tentacles shorten and thicken after attachment. Immediately after the strike, particles can be seen to pass out along the periphery of the tentacles towards the prey. Also at this time the suctorian may change shape slightly or become slightly indented (Kitching, 1952a; Hull, 1961b). Cytoplasm can now be seen to flow down the tentacles from the prey into the suctorian where it is enclosed in food vacuoles. Hull (1961b) calculated the rate of flow at about 3 μm^3/sec/tentacle. He also established that the rate of flow was constant throughout the feeding period and more or less similar for all attached tentacles. Thus animals attached by three tentacles obtain three times as much food, per unit time, than an animal attached by one tentacle. As feeding continues the areas of ciliary paralysis spread out from the point (or points) of strike until the whole surface is affected. The prey shrinks and may change shape to a sphere or appear to have been 'eaten away' at the points of attachment; contractile vacuole activity may continue for some time but eventually ceases. The contractile vacuole activity of the suctorian increases markedly during feeding (Kitching 1952b), maybe by as much as ten times as compared with the non-feeding state (Fig. 20). The function of this extra activity is to keep the increase in size of the suctorian to a minimum. Indeed the extra output may be equal in volume to half the food taken in. Finally the remains of the prey, now much reduced in size, are discarded. This last action is preceded by a movement of small granules along the tentacles from the suctorian rather like that at the onset of feeding. The tentacles appear to disengage themselves from the prey by shortening.

The electron microscope examination of suctorian tentacles (Rudzinska, 1965) showed that the tentacles, as suggested from light microscope investigation, are composed of two concentric tubes. The wall of the outer tube is the pellicle, and that of the inner tube is composed of seven sets of seven microtubules, 49 in all. This inner tube penetrates deep into the cytoplasm where it ends abruptly with an open end (Fig. 21A). At the distal end the tentacle is covered by a

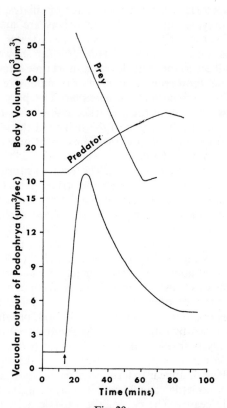

Fig. 20

Changes in vacuolar output and body volume of predator and prey during feeding in the suctorian *Podophrya*. Time of capture of prey indicated by arrow. After Kitching, 1952b (simplified).

single membrane. Protruding through this membrane are electron dense bodies that probably contain the enzymes and toxins responsible for the initial adherence and paralysis of the prey (Fig. 21B). Similar dense bodies can be seen near the base of the tentacle in non-feeding animals and during feeding they pass up the outer tube. These presumably further contribute to the digestion and paralysis of the prey. At the moment of impact with a suitable organism the knob of the tentacle is pushed into the cytoplasm of the prey, destroying the pellicle at the point of entry (Rudzinska, 1970) (Fig. 21C). The membranes of the prey fuse with those of the tentacle pellicle thus uniting the two organisms within a single membrane system. The membrane at the end of the knob stays intact and, soon after

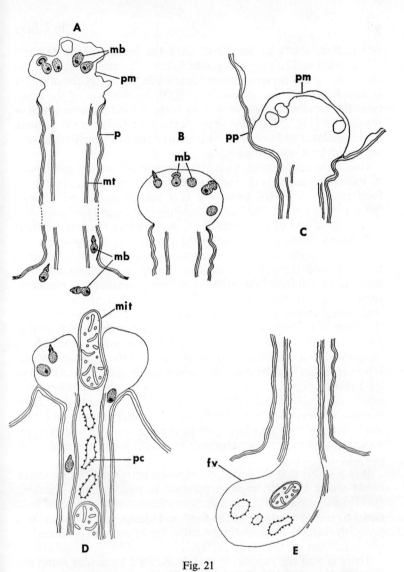

Fig. 21

Fine structure of a tentacle of *Tokophrya* before and during feeding.
A, L.S. of non-feeding tentacle. **B**, L.S. of tentacle tip showing protruding 'missile bodies'. **C**, L.S. of tentacle just after contact with the prey.
D, tentacle embedded in prey and prey cytoplasm passing down central
part of tentacle. **E**, Food vacuole formation at base of tentacle.
fv, food vacuole; mb, 'missile bodies'; mit, prey mitochondrion;
mt, microtubules of the tentacle forming boundary between inner and
outer parts of tentacle; p, alveolate pellicle of tentacle; pc, prey cytoplasm moving down tentacle; pm, plasma membrane of tentacle tip;
pp, point of fusion of the prey and tentacle pellicles. After Rudzinska
(1970).

penetration, starts to invaginate into the lumen of the tentacle (Figs. 21C,D). The prey cytoplasm follows the membrane into the invagination and is carried down the tentacle, the inner tube of which is now lined with membrane (Fig. 21D). At the base of the inner tube the membrane 'blows out' to form a food vacuole which is pinched off (Fig. 21E). Thus although there is a fusion of the pellicles of the predator and prey, the cytoplasm of the two animals is always separated by a single membrane.

One of the major problems in attempting to explain the mechanism of suctorian feeding is the location of the motive power. How is the food moved down the tentacles? Is it sucked by the predator or 'blown' by the prey? The latter is unlikely in view of Hull's observation that the rate of transport down the tentacles remains constant throughout feeding. Kitching (1952a) does not totally reject the possibility in view of the very small pressure he calculates is required to move food through the tentacles. However, Kitching and Hull arrive at very different values (0·1–1 cm and more than 200 cm water, respectively) for this pressure which must affect the interpretation of their observations. They also differ on the subject of shape change at the onset of feeding. Kitching reports dimples in the pellicle as if it had expanded. Hull did not observe these but reports a change from an ovoidal to a spherical body shape. Both of these changes could, however, result in a decrease in pressure in the suctorian. Although the two investigations are at variance over a number of points, they reach very similar conclusions: 'that both tentacles and body volume increase autonomously at the onset of feeding' (Hull) and that 'suction might well be affected by an expansion of the body surface so long as there is some resistance to a collapse of the body wall inwards' (Kitching).

It is possible that the increased vacuolar activity of the suctorian could create a slight negative pressure in the predator. Kitching (1952b) showed that the extra activity eliminated a volume of fluid equal to half the volume of the food and claims that the vacuole is probably 'helping to maintain the reduction of pressure produced by the expansion of the body surface'.

There is also the suggestion that food could be moved along the tentacles by peristalsis. Rudzinska (1965) reports periodic constrictions in the tentacle inner tube which are consistent with this idea. The wall of the inner tube is composed of microtubules, structures that are often associated with movement but there is really rather little evidence apart from the static electron micrograph pictures. In view of Rudzinska's more recent work (1970) any theory of suctorian feeding must explain the enormous increase in membrane area that results when the tentacle tip invaginates. Possibly some active

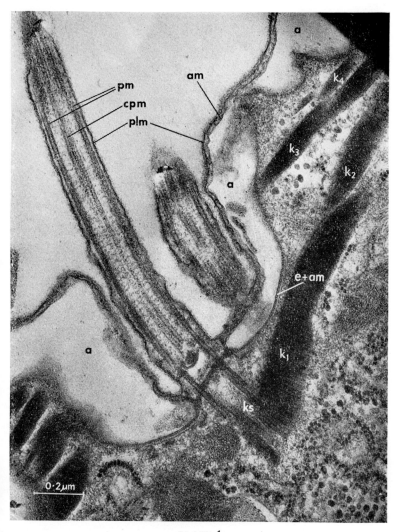

PLATE 1

Section through the cortical region of *Paramecium* at right angles to the kinety (cf. Fig. 9). The cilium is seen in longitudinal section with its peripheral microtubules (pm) and the central pair of microtubules (cpm) all enclosed within the plasma membrane (plm). The kinetosome (ks) gives rise to the kinetodesma (k_1), while the kinetodesma of other cilia can be seen near by (k_2, k_3, k_4). The alveoli (a) underlie the plasma membrane bounded by their own membrane (am) which on the inner face is elaborated into the epiplasm (e). Transmission electron micrograph kindly supplied by Dr. L. H. Bannister.

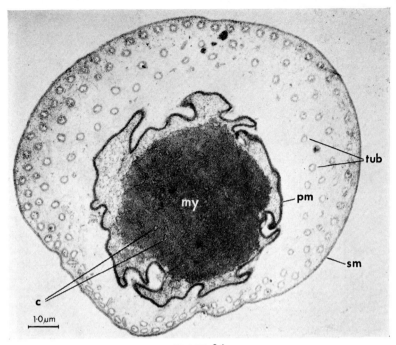

PLATE 2A

Cross-section of the stalk of the contractile peritrich *Zoothamnium* sp. showing the myoneme (my) with its canals (c), which may play a part in calcium storage (see p. 83) and plasma membrane (pm). In the annulus can be seen the supporting tubules (tub) and sheath membrane (sm). Transmission electron micrograph.

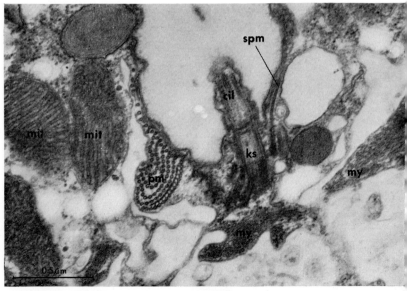

PLATE 2B

Cross-section of *Spirostomum intermedium* showing the myonemal material (my) and postciliary microtubules (pm). A cilium (cil) and kinetosome (ks) are cut in longitudinal section as are the subpellicular (transverse) microtubules (spm). mit, mitochondrion. Transmission electron micrograph.

movement of this membrane is involved in feeding. The whole problem is far from being solved.

FOOD VACUOLE FORMATION

We have seen that suctoria form food vacuoles in an extraordinary manner. The process is simpler in ciliates that are particle eaters. The particles are swept by the oral cilia into a pouch-like cytopharynx, at the base of which they are encapsulated into food vacuoles. These vacuoles circulate in the cytoplasm (Fig. 22).

What induces the formation of these food vacuoles? The two main stimuli are mechanical (presence of particles) and chemical (dissolved or particulate nutrients). It is possible to separate these stimuli by using inert particles, such as polystyrene spheres, and solution free of particles but containing dissolved nutrients. The literature is full of conflicting reports dealing with the relative effectiveness of various natural and artificial particles in stimulating food vacuole formation (Kitching, 1956b, for review). In some ciliates (e.g. *Paramecium*) particles seem to be necessary for food vacuole formation. Bozler (1924) suggests that in *Paramecium* a vacuole is initiated by the first particle to strike the end of the cytopharynx, and that the nipping off of this vacuole is stimulated by a subsequent large particle. Bozler also suggests that there might be an intrinsic rhythm of formation. He observed periodic movements of the cytoplasm at the base of the cytopharynx every 30–60 seconds even when vacuoles were not being formed.

In other ciliates vacuoles are formed in response to dissolved organic material, in the absence of particles. Seaman (1961) found that *Tetrahymena* in peptone solutions forms vacuoles and claims that protein molecules are the stimulus. He found that *Tetrahymena* in solutions of aminoacids did not form vacuoles. Thus, he claims, food vacuole formation can be induced in a similar way to pinocytosis in amoebae. However, aminoacids and salts, both active inducers of pinocytosis, do not stimulate food vacuole formation so that the parallel is far from exact. Holz (1964) disputes the necessity for protein inducers, reporting that vacuoles can be seen in animals repeatedly washed and then left to stand in distilled water, but admits that both particles and proteins are more active inducers than small molecules or non-nutrient media. Recently Rasmussen and Modeweg-Hansen (1973) have shown that reproduction of *Tetrahymena* is almost absent if the medium is filtered of all its particulate matter. The addition to the medium of a variety of inert particles results in increased growth. This suggests that particles stimulate food vacuole

D

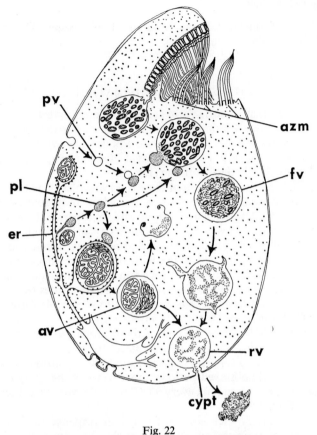

Fig. 22

Ingestion and digestion in *Tetrahymena*

Food vacuoles are formed at the base of the cytopharynx. Primary lysosomes containing digestive enzymes fuse with the food vacuoles. When digestion is almost complete the products are enclosed in small vacuoles that move into the cytoplasm, while the indigestible remains are voided to the exterior through the cytoproct. The animal may also, in periods of starvation, digest its own organelles in autophagic vacuoles. av, autophagic vacuole; azm, adoral zone of membranelles; cypt, cytoproct; er, endoplasmic reticulum; fv, food vacuole; pl, primary lysosome; pv, pinocytic vesicle; rv, vacuole containing undigested residue. After Elliott and Clemmons (1966).

formation which in turn are important in the uptake of soluble nutrients.

Ciliates appear to have some selective ability with regard to the particles that they take into their food vacuoles. Kitching's review mentions a number of independent observations that particles such as bacteria, powdered egg yolk and yeast cells are taken up preferentially to sulphur or powdered glass. However, in the absence of the former non-nutrient particles will be taken up (i.e. they are not absolutely discriminated against). Much work has been done on this selection process but no clear evidence in support of discrimination by chemical means has appeared; rather, selection seems to be based on the physical properties of size and density, e.g. *Paramecium* will reject large particles. Bozler (1924) showed that sepia particles were more readily accepted than carmine unless the latter was ground as finely as the sepia in which case both were taken in equal quantities. Bozler was also of the opinion that the shape of particles affect their acceptability, the sharp angularity of many inert particles accounting for their unattractiveness. Rapport *et al.* (1972) were able to show that *Stentor coeruleus* does have food preferences. They removed the variable of 'catchability' due to size by presenting various prey species alone and in pairs and comparing consumption under the two conditions. They showed a general preference for non-algal over algal food. Müller *et al.* (1965) investigated the uptake of polystyrene latex particles (PLP) by *Paramecium* and *Tetrahymena*. He found that PLP were as acceptable as the bacterium *Aerobacter*. There was no difference in the frequency of food vacuole formation in PLP and bacterial suspensions.

DIGESTION IN FOOD VACUOLES

Food vacuoles once formed move about in the cytoplasm until the digestion of the contents is completed. There are four stages in the life of such a food vacuole.

(1) The vacuole is nipped off at the base of the cytopharynx and for a short time it, and its contents, remain virtually unchanged, except for a reduction in size. There is no enzymic activity associated with the vacuole.

(2) The reduction in size continues and there is a sharp fall in the pH of the vacuolar fluid. The vacuole becomes surrounded by granules called neutral red granules by the light microscopist because of their staining properties. In electron microscope preparations they have the appearance of lysosomes. These granules contain digestive enzymes that are passed into the vacuoles where their activity can be

detected. Müller *et al.* (1965) studied the distribution of the digestive enzyme, acid phosphatase, during the feeding. They found that just after feeding *Paramecium* (with PLP, so that enzymes in the food could not interfere with their studies) there was no enzymic activity in or near the vacuole. After 2 or 3 minutes, however, granules appeared around the wall of the vacuole which had strong enzymic action. A short time later the enzyme was to be found in the vacuole wall itself. Unfed animals had such enzyme granules scattered throughout the cytoplasm. Comparisons of fed and unfed animals suggested that there was a redistribution (rather than a change in quantity) of acid phosphatase after feeding. Elliot and Clemmons (1966) performed similar experiments, feeding *Tetrahymena* (Fig. 22) with mixtures of bacteria and india ink particles and studying the acid phosphatase distribution by electron microscope histochemistry. In this organism the first stage lasts as long as 30 min. After this strongly staining vesicles appear close to the vacuole and apparently discharge their contents into the vacuole. At this time there is a strong reaction from the enzyme within the vacuole and the bacterial food appears partially digested. A short time later little remains of the bacteria and the enzyme activity can only be found near the now irregular wall of the vacuole. The enzymic granules (primary lysosomes) are formed in the rough endoplasmic reticulum (RER). After their formation they fuse with vacuoles produced by pinocytosis at the pellicle. After this fusion the granules pass to the food vacuole. Estévé (1970) has shown that the acid phosphatase manufactured in the RER is packaged in a Golgi apparatus as in many metazoan cells.

The electron microscope study of food vacuoles and digestion in *Colpoda* (Rudzinska *et al.*, 1966) is in general agreement with other studies. The beginning of the second stage is heralded by a very marked shrinkage of the vacuole, so great that the bacterial food is squashed and distorted. This state is presumably the result of the withdrawal of water from the vacuole and ensures least dilution of digestive enzymes. Endoplasmic vesicles are to be found close to the food vacuole at this stage and may well be the source of digestive enzymes.

(3) The products of digestion now leave the vacuole by a process of micropinocytosis. Small vesicles form in the wall of the vacuole and pass into the cytoplasm. Enzymic activity can be detected in these vesicles, and it is possible that digestion continues. At this time the vacuole enlarges somewhat.

(4) In the final stage the vacuole contains indigestible remains; mainly membranous material in the case of bacterial food. Occasionally almost intact bacteria may be found in such vacuoles sug-

gesting that newly formed vacuoles may fuse with spent ones. The contents of a spent vacuole are ejected into the environment, usually through a fixed spot on the pellicle, known as the cytoproct.

Ciliates commonly consume their own organelles in the absence of food. Digestion is carried out in food vacuoles which appear very like normal food vacuoles. Such autophagic vacuoles may contain membrane debris, etc., and seem to be supplied with enzymes by primary lysosomes.

The normal series of events described above are paralleled by changes of acidity within the vacuole (Kitching, 1956b). These changes can be detected by feeding animal starch grains or similar food which has been impregnated with pH-sensitive dyes, or by putting the dyes in the medium so that a certain amount is taken up with the food. At first the pH is the same as that of the medium. However, during the digestive phase the pH falls. Typically values of pH3·2–4·0 have been obtained but values as low as pH 1·0–1·4 have been recorded. It seems likely that such changes are due to acid secreted by the ciliate.

Relatively little is known about the exact nature of the digestive enzymes (Müller, 1967). Acid phosphatase has often been studied because a convenient technique is available for its localization. However, as Müller points out, the demonstration of the presence and activity of enzymes is only the first step in understanding intracellular digestion. Although a great many enzymes have been identified in ciliates there is little evidence as to their role in digestion.

Food vacuoles have been considered in the context of the wider field of lysosome studies (Novikoff, 1961) embodying work on cells of metazoa. Lysosomes are characterized by having a set of hydrolases that will split almost all types of biological molecules. The surrounding membrane does not let the enzymes out or other substances in. Primary lysosomes are usually formed in the RER and Golgi apparatus. Secondary lysosomes are formed by a fusion of autophagic or endocytic vacuoles with lysosomes (usually primary). Occasionally secondary lysosomes fuse together. The products of enzyme action may be removed from a secondary lysosome into the cytoplasm. Clearly the food vacuoles and associated vesicles of ciliates qualify as lysosomes according to these descriptions, and it may be rewarding to view ciliate digestion within this wider context (Müller et al., 1963).

SYMBIOSIS

Some ciliates have symbiotic green algae living in their cytoplasm. Twelve such genera of ciliates are listed by McLaughlin and Zahl

(1966) of which the commonest species are *Stentor polymorphus*
and *Paramecium bursaria*. In both of these animals the endosymbionts are green zoochlorellae. In both cases the ciliate derives some
nutritional benefit from the association. Pringsheim (1928) placed
P. bursaria in sealed vessels, thus limiting their oxygen supply. Those
that had been deprived of their zoochlorellae soon died while
ciliates with them survived for months. Those in continuous light
survived best of all. Pringsheim claimed that the endosymbionts
were supplying the ciliates with oxygen produced in photosynthesis.
MacLaughlin and Zahl (1966) claim that photosynthetically produced oxygen is of no importance but this may not be true in habitats where the oxygen tension is low. Pringsheim also claimed that
ciliates with zoochlorellae could be maintained indefinitely in sterile
fluid containing only inorganic materials, indicating that the endosymbionts could supply all the ciliates', and their own, nutritional
needs by photosynthesis. Loefer (1936) could not confirm this claim.
However, there is no doubt that the zoochlorellae supply *P. bursaria*
with some nutrients (Karakashian, 1963). Equally when bacterial
food is plentiful the ciliate can support the algae in the absence of
light. Muscatine *et al.* (1967) have shown that algae isolated from
P. bursaria and grown axenically excrete from 5–68 per cent of their
photosynthate (mostly maltose) into the medium as compared with a
maximum of 7·4 per cent for free-living *Chorella* species. If such loss
occurs inside the ciliate the nutritional advantage to the latter is
obvious. It is interesting to note that zoochlorellae isolated some 35
years earlier showed less excretion of sugars than those recently
isolated. Other substances, such as vitamins, may also be supplied
by the zoochlorellae. The flagellate *Crithidia oncopelti* appears to
require only a single aminoacid, in contrast to the dozen or so needed
by its relatives. Newton (1957) showed that these flagellates contain
bacterial endosymbionts which probably supply their hosts with the
other aminoacids that they need. Bacterial endosymbionts occur in
various ciliates including species of *Paramecium* and may play a
similar role in supplying proteins or aminoacids. There is some evidence that the bacterial endosymbiont, kappa, may contribute to
synthesis of the purine, guanine, in *Paramecium*.

   It is also likely that when the ciliates are short of food they consume their endosymbionts. Karakashian *et al.* (1968) frequently
found algae in the food vacuoles of *P. bursaria,* but it is possible that
these cells may have been taken in from the medium. They maintain
that digestion of zoochlorellae may be a normal part of the symbiotic
relationship. When *Stentor polymorphus* is starved it becomes colourless, although Tartar (1961) claims it is difficult to rid *Stentor*
completely of its symbionts in this way.

It is possible to infect ciliates with algae from other host species; some establish stable symbioses, some are rejected (Tartar, 1961; Karakashian and Seigel, 1965). The success of any given cross infection is not predictable, depending not upon any given property of ciliate or algae, but upon properties peculiar to the particular combination. Karakashian and Seigel point out that one ought to regard the association as having a separate genotype from either of its members. They cite evidence (though not from ciliates) suggesting that symbiotic associations can produce chemicals which neither of the members alone can produce. Symbiosis in this way can be producing something more than a simple summation of the capabilities of the partners.

SAPROZOIC NUTRITION

Many ciliates are able to utilize nutrients dissolved in their environment. Indeed some are able to grow and multiply in the absence of particulate food, providing that all the essential nutrients are available in the medium. This allowed cultures to be established which were free of bacteria. Such axenic cultures have now been achieved for many species. Early axenic culture media contained complex organic materials (e.g. yeast or meat extracts, milk, etc.). However, in some cases it has been possible to substitute specific chemicals for these crude undefined mixtures. By a process of trial and error, unessential substances can be eliminated until one finally arrives at a medium that represents the basal requirements. The nature of such requirements sheds considerable light on the biochemistry of the animals concerned. Although many ciliate species can be grown in defined or semi-defined media, most of the research effort has concentrated on one species: *Tetrahymena pyriformis*.

Aminoacids and peptides

In general it seems that ciliates like many other animal cells require the aminoacids: argenine, histidine, leucine, isoleucine, lysine, methionine, phenylalanine, threonine, tryptophan and valine. Some species or strains have additional requirements, e.g. proline, tyrosine or glycine (see Holz, 1964, for review). *Paramecium aurelia* has an absolute requirement for 12 aminoacids, but growth is optimum in the presence of 13 or 14. In some cases one or more aminoacids can substitute for another, e.g. serine can replace glycine but only in the presence of thymidine (Soldo and Van Wagtendonk, 1969).

Many ciliates grow better if they receive their aminoacids in the form of proteins or peptides; indeed for some time there was thought to be requirement for protein. Holz is of the opinion that proteins stimulate growth because they may (1) correct aminoacid imbalances in the medium, (2) stimulate food vacuole formation, (3) supply, as contaminants, unrecognized growth stimulants (usually lipids), (4) bind toxic materials and (5) supply essential aminoacid sequences. He adds that 'the dreary history of "protein requirement" in microbial nutrition favours every interpretation but the last'.

Some ciliates secrete protolytic enzymes into the environment. At first sight this would appear to be advantageous in that proteins in the medium would be converted into more readily absorbed peptides and aminoacids. However, Müller (1967) rightly points out that in the natural environment the enzymes would suffer such dilution as to be useless to their producers.

Purines and pyrimidines

Ciliates have RNA and DNA which contain the purines guanine and adenine, and the pyrimidines uracil, cytidine and thymidine. These organic bases must be either synthesized or obtained from the environment. *Tetrahymena pyriformis* is able to convert guanine to adenine but is unable to perform the reverse reaction, and guanine is therefore an absolute requirement (Flavin and Graff, 1951). *T. paravorax* is able to grow when supplied with guanylic acid as its only purine source. *Paramecium* with kappa particles (see p. 102) is able to synthesise guanine from adenine, but is less efficient at this synthesis after losing its kappa particles.

The pathways for pyrimidine synthesis seem a little more open than those for purines and usually any one of a number will suffice. For instance *Paramecium aurelia* can manufacture its pyrimidines from uridine, uridylic acid, cytidine or cytodylic acid and *Tetrahymena pyriformis* needs either uracil or cytidine.

Carbon sources

A wide variety of carbohydrates, organic acids and alcohols will act as carbon and energy sources for ciliates. Of the carbohydrates glucose seems to be the most generally useful but some species do better with dextrin. Glycogen and starch can also be utilized, but sucrose is often of limited use. Some ciliates found in the rumen of cattle are able to utilize various plant carbohydrates including cellulose, hemicellulose, starch, pectin and fructan. It is not yet established that the ciliates possess a cellulase as those that occur may be produced by the rumen bacteria. Both acetate and pyruvate can be used by many ciliates, especially species of *Paramecium*. *Tetrahymena setifera* in defined medium has an absolute requirement for either methanol or ethanol (Holz *et al.,* 1962): an alcoholic among ciliates! Another biochemical curiosity is to be found in *Tetrahymena pyriformis* which is able to fix carbon dioxide under anaerobic conditions (Ryley, 1952; 1967).

Many ciliates form carbohydrate stores in times of plenty. *Tetrahymena* forms glycogen under aerobic conditions. Some rumen ciliates are able to synthesize amylopectin, one of the constituents of typical plant starch. Indeed this substance may constitute up to 70 per cent of the cells' dry weight and Sugden and Oxford (1952) report that in the presence of high glucose concentrations the ciliates may form so much of it that they burst.

Only in *Tetrahymena* has the biochemistry of respiration been examined in any detail. Both glycolosis and the hexose monophosphate shunt seem to be available for the production of pyruvate from polysaccharides and hexoses. The enzymes for both pathways have been demonstrated but there is no evidence to suggest which is the more important under natural conditions. Under aerobic conditions pyruvate would pass into the tricarboxylic acid (TCA) cycle which appears to be present in *Tetrahymena* in a complete form (Danforth, 1967). Most of the important enzymes of the cycle have been demonstrated in this animal and it is able to oxidize most of the TCA cycle acids. The ciliate possesses enzymes to process amino and fatty acids for entry into the TCA cycle and indeed *Tetrahymena* does not metabolize carbohydrate much under aerobic conditions and the indications are that proteins and fatty acids are the normal respiratory substrates (Ryley, 1967). There is some dispute as to whether or not *Tetrahymena* has a complete cytochrome chain. Cyanide does inhibit respiration but it is not clear that the site of action is cytochromoxidase as it is in mammalian systems. Under anaerobic conditions endogenous (glycogen) and exogenous carbohydrates are fermented to lactic acid, if carbon dioxide is present, or to succinate, if it is absent. Despite some peculiarities the aerobic respiratory biochemistry of *Tetrahymena* is broadly similar to that of other animals.

Growth factors and vitamins

Ciliates require various complex organic compounds at very low concentrations (see Lilly, 1967, for review). Thiamine (vitamin B_1) is a commonly required vitamin for protozoa in axenic culture. It has been shown for *Paramecium* and for all but a few mutant strains of *Tetrahymena*. Its principal role is in the formation of a coenzyme involved in the decarboxylation of pyruvate. Riboflavin is also required by many ciliates investigated. Vitamin B_6 may be supplied in three forms; pyridoxin, pyridoxal and pyridoxamine. In higher animals these three forms are equipotent but many strains of *Tetrahymena* cannot utilize pyridoxin and require one of the other two. All ciliates studied require at least one form of the vitamin.

Ciliates have also been shown to require pantothenic acid (vitamin B₅), nicotinamide (vitamin B₃) and folic acid or some other pterdine moiety. Thioctic acid has been shown to be essential for the growth of tetrahymenids, but not for other ciliates.

Lipids

The defined media of a number of species of ciliates contain lipids (see Dewey, 1967, for review). However, it is not always clear whether these are absolute requirements for growth. In some cases the requirement has been established. For example *Paramecium multimicronucleatum* requires a sterol, of which stigmasterol is the most effective. Various species of *Tetrahymena* require a sterol. Some can can use cholesterol while others can utilize the precursor, demosterol.

BIOLOGICAL ASSAY WITH CILIATES

If a certain substance, e.g. an aminoacid, is an absolute requirement for growth, it is often the case that, provided the other nutrients are available in optimum amounts, the rate of growth will be proportional to the concentration of that substance. This will hold only over a certain concentration range. Below certain levels there may be no growth while at a certain higher concentration, further addition does not enhance the growth rate (indeed it may be inhibitory). Between these concentrations growth is proportional to the amount of nutrient present. Having established the relationship between concentration and the rate of growth with known standard media, it is then possible to establish the concentration of that nutrient in 'unknowns'. Many micro-organisms, including ciliates, are used in this way to assay certain organic chemicals. *Tetrahymena pyriformis* has been used to determine concentration of pantothenic acid. It is sensitive over a wide range (1–300 ng/ml, Baker *et al.*, 1960). The method can be used to measure pantothenic acid levels in human subjects. The serum is added to media which lacks only this vitamin and the growth is compared with that in media containing various known concentrations of pantothenic acid. *T. pyriformis* has also been used to assay nicotinic and thioctic acids (Hall, 1965).

CILIATES IN BIOCHEMISTRY

Ciliates are widely used in biochemical investigations, both as tools and for their own sake. This work is largely beyond the scope of this book and the reader is referred to the reviews by Kidder and Dewey (1951), Seaman (1955, 1961), Holz (1964), Connor (1967), Dewey (1967), Kidder (1967), Lilly (1967), Mandel (1967) and Ryley (1967).

5

OSMOTIC AND IONIC REGULATION

The cytoplasm of ciliates is separated from the environment by a delicate and relatively thin series of membranes (see p. 40). The composition of the cytoplasm and the concentration of its constituents are different from the outside medium and thus the animals are subjected to various ionic and osmotic stresses. These stresses are least in marine forms and greatest in freshwater ones. The latter may be maintaining an internal concentration equivalent to 50 mM of sucrose (Kitching, 1967) in an environment perhaps as dilute as 5 mM. Differences also exist in the ratios of various ions, inside and outside the animal; internal sodium is often a smaller proportion of the total ions as compared with the medium, while potassium inside fills a larger proportion than outside (Carter, 1957; Dunham and Child, 1961; Dunham, 1969; Hilden, 1970). A freshwater ciliate may be simultaneously subjected to entry of water by osmosis, and sodium by diffusion, while losing potassium by diffusion. All these movements must be controlled and compensated for if the cytoplasm is to retain any degree of constancy.

CONTRACTILE VACUOLES AND OSMOREGULATION

The morphology of the contractile vacuoles has been considered above (p. 38). They are found in all freshwater ciliates and, contrary to popular belief, in many marine and parasitic species also (Kitching, 1938). In marine species the vacuolar duration (the time from one systole to the next) is long compared with that of freshwater forms. Vacuolar duration need not, however, represent the output of the vacuole. A true figure for output can be obtained by multiplying the average volume of the vacuole just before systole by the average vacuolar duration. This value can now be compared with the animal's volume or surface area. For example the marine peritrich *Cothurnia curvula* eliminates a volume of vacuolar fluid equal to its own volume in 4·5 hours (Kitching, 1938), while the freshwater *Halteria grandinella* clears its own volume in only 7·4 minutes. Other examples can be found below.

Rate of vacuolar output (Kitching, 1938)

Species	Time required to eliminate quantity of water equal in volume to body
Freshwater ciliates	
Paramecium caudatum	14–49 min
P. aurelia	46 min
Lembus pusillus	4·1 min
Urotricha globosa	12 min
Cyclidium glaucoma	6·6 min
Urocentrum turbo	13·2 min
Colpidium colpoda	53 min
Frontoma leucas	260 min
Halteria grandinella	7·4 min
Stylonichia pustulata	20·5 min
S. mytilus	45 min
Euplotes patella	14·3 min
Vorticella convallaria	47·3 min
Zoothamnium sp.?	24·5 min
Marine ciliates	
Cothurnia curvula	4 hr 15 min
C. socialis	4 hr 45 min
Zoothamnium marinum	2 hr 45 min
Z. niveum	3 hr 15 min
Endoparasitic ciliates	
Nyctotherus cordiformis	4 hr 5 min

What is the source of the fluid expelled by the contractile vacuole? There are three possibilities. (1) Water taken in with the food vacuoles during feeding. Some workers have considered this an important source of fluid as the food vacuole can be observed to shrink soon after formation. However, the spent food vacuole increases in volume and when its contents are voided through the cytoproct its size is similar to that at formation. The net gain of fluid to the animal is therefore very small. Kitching (1967) has pointed out that ciliates with no mouths, and which do not form food vacuoles, and those that feed intermittently have continuously active contractile vacuoles. (2) Water derived from metabolism. This can be estimated, assuming that the consumption of 1 ml of oxygen gives rise to 1 ml of water, as is the case with the breakdown of glucose. Kitching (1956a) calculated that *Paramecium caudatum* produces about 2000 μm^3 of metabolic water per organism per hour, which is less than 1 per cent of the vacuolar output. (3) Water entering by osmosis. If one assumes that the ciliate pellicle is selectively permeable and that the internal concentration exceeds the external concentration (which is true of freshwater species), then water will pass into the animal by osmosis. The animal must then either swell until it bursts or become diluted until it is isosmotic with the external fluid, or it may resist swelling mechanically as a plant cell does. The final possibility is to bale out the water as fast as it enters. This last seems to be the function of the contractile vacuole. The major piece of evidence supporting an osmoregulatory role is that when the external medium is diluted, and osmotic inflow presumably increased, the output of the vacuole increases. Fig. 23 shows the relation of output to external concentration

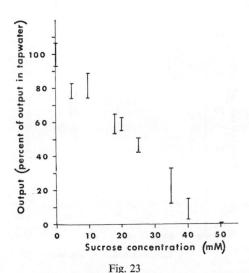

Fig. 23

The vacuolar output of *Podophrya collini* in relation to the sucrose concentration in the medium (output with no sucrose taken as 100). Simplified from Kitching (1967).

in *Podophrya collini*. The straight line crosses the abscissa at 50 mM
sucrose indicating that normally the osmotic difference between the
animal and its environment is equivalent to about this concentration
of sucrose. This suggestion is strengthened further by the observation that if a freshwater peritrich is poisoned with cyanide, which
inhibits the contractile vacuole, the animal just fails to swell in a
50 mM sucrose solution (Kitching, 1938).

Although there can be little doubt that the vacuolar fluid of freshwater ciliates derives from osmotically entering water the relation of
output to external concentration is not always linear. This is especially true of marine ciliates such as *Cothurnia*. In this case initial
dilution of the sea-water has little effect on output, but below 50 per
cent sea-water output rises steeply only to fall again at very low external concentrations. It is difficult if not impossible to decide on the
internal concentration of *Cothurnia* from such a graph (Fig. 24).
However, there is a certain similarity between this curve and figures
obtained for osmoregulation of some brackish water invertebrates
which suggests the following interpretation. *Cothurnia* loses salts if

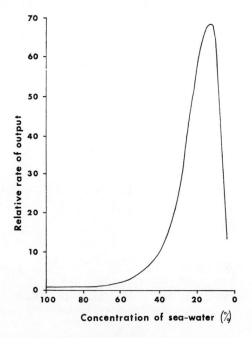

Fig. 24
The vacuolar output of *Cothurnia curvula* in various concentrations of
sea-water. Simplified from Kitching (1934).

the external fluid is diluted, remaining isotonic with the environment down to about 50 per cent sea-water. Upon further dilution of the medium salts are retained and osmotic water entry increases. The vacuole responds with a higher output. At greatest dilution the animal is unable to retain salts which leak out leading to a reduction in the entry of water.

CONCENTRATION OF CILIATE CYTOPLASM

It is assumed above that the internal concentration of freshwater ciliates is above that of the external medium. What is the evidence for this? Picken (1936) showed by vapour pressure measurement that *Spirostomum* protoplasm has a concentration equivalent to 25 mM/l of sodium chloride, although his method was not altogether reliable. Carter (1957) found by isotopic equilibration that the total of internal sodium, potassium and bromine concentrations varies from 33–43 mM/l depending on the external concentration, which it exceeds even at 41 mM/l. *Tetrahymena* (Dunham and Child, 1961; Andrus and Giese, 1963) and *Blepharisma* (Hilden, 1970) are probably hypertonic to the environment especially when the latter is dilute. Internal calcium is in excess of external concentrations in *Paramecium* (Yamaguchi, 1960) and *Spirostomum* (Jones, 1966) but at least in the case of *Spirostomum* this is osmotically inactive because as much as 90 per cent of it is bound (Jones, 1967). The organic molecules of the ciliate cytoplasm must also contribute to their hypertonicity.

As already stated the behaviour of the contractile vacuole suggests that freshwater ciliates are hypertonic to the medium. However, if one assumes a hypertonic cytoplasm to explain contractile vacuole behaviour it is a circular argument to then cite vacuolar behaviour in support of hypertonicity. This circularity can be broken if other evidence can be produced to show that the vacuole is baling out water. Such evidence is supplied by a study of volume regulation.

CONTROL OF BODY VOLUME

The contractile vacuole is often thought of as being analogous to a kidney, controlling the internal concentration of the animal by expelling excess water. However, in certain circumstances its function seems more that of controlling body volume rather than concentration. Kitching (1948b) found that if the peritrich *Carchesium aselli*, cultured in pond-water, was placed in pond-water to which ethylene glycol had been added the vacuolar behaviour was as follows. At first the output fell markedly in proportion to the concentration of the ethylene glycol, but after a few minutes it began to rise again (Fig. 25) and eventually reached the same value as in pond-water alone. The animal was then returned to pond-water and vacuolar output rose markedly, but soon fell to the original pond-water value. The stronger the ethylene glycol the more exaggerated were the output changes. This experiment can be interpreted as follows. Upon initial immersion in the ethylene glycol solution, the increase in external concentration reduces the osmotic inflow and the vacuolar output is reduced in response. If the concentration of ethylene glycol is sufficient the output may reach zero for some time. Ethylene glycol

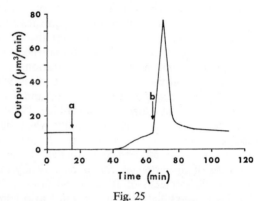

Fig. 25

The vacuolar output of *Carchesium aselli* when subjected to pond-water containing 250 mM ethylene glycol. a, animal transferred from pond-water to pond-water containing ethylene glycol; b, returned to pond-water. Simplified after Kitching (1956a).

gradually diffuses into the animal, increasing its osmotic pressure and hence the influx of water and again the vacuole responds. Eventually the ethylene glycol is at an equal concentration on both sides of the ciliate pellicle and the concentration difference between the inside and the outside is the same as in pond-water alone. At this point the peritrich, with an internal concentration raised by a quantity of ethylene glycol, is returned to pond-water. The additional solute results in a massive inflow of water by osmosis with an associated rise in vacuolar activity, but the ethylene glycol quite quickly diffuses out into the medium restoring the initial situation. During the course of this experiment the contractile vacuole has been active in preventing excessive swelling of the peritrich and does not seem to respond to the considerable changes in the internal concentration, although it is possible that ethylene glycol excretion may occur through the vacuole.

Podophrya is also subjected to changes in body volume during feeding (see p. 93). Kitching (1952b) found that the activity of the contractile vacuole increased during suctorian feeding (Fig. 20). He drew up a balance sheet setting the increase in volume of the *Podophrya* plus the extra amount of water pumped out against the decrease in volume of the prey. He obtained good agreement between the two sides of the sheet. One typical animal increased in volume by $5 \times 10^3 \, \mu m^3$ and the water expulsion increased by $3 \cdot 1 \times 10^3 \, \mu m^3$ while the prey decreased by $7 \cdot 8 \times 10 \, \mu m^3$. As can be seen in Fig. 20 the contractile vacuole activity quickly responds at the onset of feeding though the output gradually falls off as feeding progresses. Kitching was able to show that the increased activity was not due to an increased internal osmotic pressure. The output after feeding is at a new steady level greater than before feeding. This may be due to an increased internal osmotic pressure due to the food which it has taken in and which has been concentrated by the action of the contractile vacuole. It is also possible that the increased surface area after feeding results in a higher inflow of water. There is evidence that the permeability of the ciliate pellicle increases during and after feeding (Carter, 1957; Jones, 1966) and this may contribute to increased water inflow.

Here again we see the contractile vacuole acting not to regulate internal osmotic pressure, but to regulate or at least to minimize changes in body volume. There is an obvious advantage to suctorians in being able to concentrate their food and so allow room for more. Kitching showed that more than a third of the volume of the food taken in was disposed of via the contractile vacuole.

Stoner and Dunham (1970) investigated the internal concentration of sodium, potassium, chloride and aminoacids in *Tetrahymena* following the addition of 100 mM sucrose to the external medium. Initially the animals shrunk to about two-thirds of their original volume. The vacuolar activity ceased. Gradually the internal concentration of sodium and aminoacids rose increasing the osmotic pressure of the animal and therefore the water entry. This influx combined with continued vacuolar inactivity resulted in a gradual return to normal volume. When this is restored the contractile vacuole resumes activity. In this example there is a complex interaction between internal concentration, body volume and vacuolar function.

Body volume also plays an important role in controlling the contractile vacuole of the marine peritrich *Cothurnia* (Kitching, 1936). When this animal was transferred from 100 per cent to 12·5 per cent sea-water the body volume increased sharply with an associated rise in vacuolar output. As salts diffuse out of the ciliate the animal's volume gradually decreases, vacuolar activity falling as well. On

return to 100 per cent sea-water both body volume and activity fall
to values lower than those first observed in this medium until salts
diffuse back to restore the original state.

EFFECTS OF TEMPERATURE ON THE CONTRACTILE VACUOLE

Kitching (1948a) showed that the vacuolar activity of *Carchesium*
varied with temperature. He subjected this ciliate to various temperatures and compared the output with that at 15°C. Fig. 26 shows his

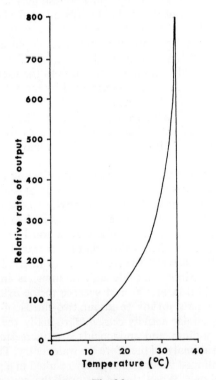

Fig. 26
Relation between vacuolar output and temperature in *Carchesium
aselli*. After Kitching (1948a).

results. At near freezing point the maximum diameter of the vacuole
was large but the very long interval between systoles resulted in a
low output. At temperatures above 15°C there was an increase in the
frequency of systole and at the highest temperatures there is an
increase in the maximum diameter. At 34°C the vacuole could only

maintain a high output for a short time before the frequency dropped and the animal swelled and died.

The Q_{10} for output is about 2·5–3·2. This is higher than can be explained by the effect of temperature on the osmotic pressure of the animal (Q_{10} about 1·05). It could be that rise in temperature stimulates the contractile vacuole to increased activity without there being much increase in the entry of water. If this is the case then the animal should shrink. Kitching calculated that this shrinkage, at about 30°C, would be in the order of 50 per cent in 15 minutes. As he failed to observe any such volume change he assumes that the increased activity reflects an increase in water influx. Temperature could raise the osmotic pressure of the animal by stimulating the production of metabolites or it could increase the permeability of the pellicle. However, Kitching (1948b) showed that the former is unlikely to be true as the osmotic pressure is equivalent to 60 mM sucrose at both 15 and 30°C. He produced evidence in support of increased pellicle permeability by subjecting the marine peritrich, *Vorticella marina,* to dilute sea-water. These animals swell under such circumstances, the rate of swelling being greater at high temperatures and having a Q_{10} of about 2·5–3·2.

Kitching (1954a) further showed that in *Podophrya* a rapid rise in temperature induces an initial large rise in the frequency of systole which then falls to the level characteristic for the temperature. The reverse is true with a rapid drop in temperature; first a great reduction in the frequency of systole followed by a gradual rise to a steady level.

EFFECT OF OTHER CHEMICAL AND PHYSICAL FACTORS ON VACUOLAR ACTIVITY

Pressure affects contractile vacuole activity. Kitching (1954b) showed that at 2000–3000 psi (pounds per square inch) the frequency of systole increased, but decreased at 5000 psi and at 7000 psi it was completely suppressed. It is possible that the effects of 2000 psi could be interpreted as an instability of the plug blocking the vacuole pore or as a stimulation of the contractile sol–gel changes in the surrounding cytoplasm. At higher pressures metabolic processes are inhibited leading at 7000 psi to complete inhibition of vacuolar activity.

Czarska (1964) claims that the contractile vacuole of *Paramecium* responds to electrical stimulation. *Paramecium* has two contractile vacuoles and Czarska states that during continuous d.c. stimulation the vacuole nearest the anode increases its activity whilst that nearest the cathode has a reduced activity. However, these results are very difficult to interpret in view of the profound effects that electrical stimulation has upon ciliates. Czarska also found that, as compared with glucose or urea, potassium chloride was more effective in reducing the frequency of systole. Mixtures of calcium and potassium chlorides with similar Gibbs–Donnan ratios (see p. 77) caused almost no change in frequency even if their total osmolarities were very different. She sees this last result as evidence for a similarity between cilia and the contractile vacuole, and suggests this supports the 'contractile' nature of the vacuole. This work needs careful reinvestigation.

The output of the contractile vacuole can be reversibly inhibited by concentrations of cyanide as low as 10^{-5} M (Kitching, 1936), there being a drop in the final diameter of the vacuole as well as in the frequency of systole. At high concentrations of cyanide there is a complete cessation of vacuolar activity. Kitching used cyanide inhibition to estimate internal osmotic pressure. He stopped the vacuole with cyanide and then found the lowest concentration of sucrose in which the animal did not swell. This concentration was taken as being equivalent to the internal concentration.

Extracellular application of adenosine triphosphate (ATP) increases the frequency of systole (Organ et al., 1968b). These workers showed that the control Paramecium had an average of 3·68 systoles per minute compared with 6·25 for animals treated with 3·0 mM ATP. It was assumed that the effect was not due to increased permeability as shrinkage of the cells resulted from this increased activity.

IONIC REGULATION

It is apparent from a number of studies that ciliates are not only able to regulate their volume, and to some extent their internal osmotic pressure, but also the concentration of their ions. Freshwater ciliates tend to lose salts by diffusion both through the pellicle and in the fluid excreted by the contractile vacuole. There are no accurate figures available for the concentration of vacuolar fluid of ciliates but Schmidt–Neilsen and Schrauger (1963) have obtained freezing-point depression determinations from Amoeba proteus. In this case the vacuolar fluid was 32 mOM compared with values of 101 mOM for the cytoplasm and 8 mOM for the medium. It is reasonable to suppose that similar figures pertain for freshwater ciliates. The loss of salts through vacuolar fluid must be made good either in the diet or by the uptake of salts through the pellicle.

The pellicle of ciliates appears to be permeable to ions. Dunham and Child (1961) found that the times for maximum exchange of radioactive sodium and potassium in Tetrahymena were 120 min and 180 min respectively and that half the exchange took place in 3 min and 30 min respectively. Despite this permeability of the membranes the internal concentrations of ions are often different from those of the external medium (Fig. 27). Changes in the external concentrations of ions do not always lead to similar changes within the animal. For

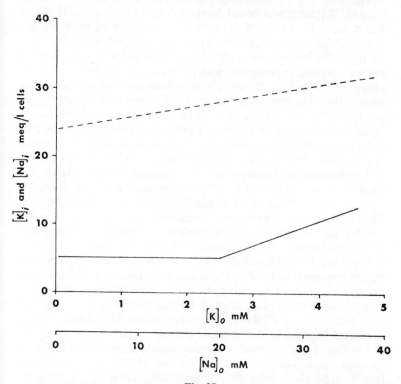

Fig. 27

Relation between internal and external concentrations of sodium (unbroken line) and potassium (dashed line) in *Tetrahymena pyriformis*. After Dunham and Child (1961).

instance, Dunham and Child found that in normal medium the internal potassium concentration ($[K^+]_i$) was 31.7 m eq/l of cells compared with an external concentration ($[K^+]_o$) of 4.75 mM/l. Internal sodium ($[Na^+]_i$) was 12.7 m eq/l compared with the external concentration ($[Na^+]_o$) of 36.5 mM/l. They placed *Tetrahymena* from normal medium into one which had been diluted up to 100 fold. Small falls in $[Na^+]_i$ and $[K^+]_i$ were observed. The $[Na^+]_i$ drop occurred soon after dilution whereas the fall in $[K^+]_i$ was slower and more gradual. Cells subjected to a 20-fold dilution for 2 days lost half their internal potassium. There was no significant change in cell volume with 10-fold dilution of the medium so presumably the contractile vacuole was able to cope with the increased water influx.

The response to increasing the external concentrations was as follows. If $[Na^+]_o$ was raised from 36·5 mM/l to 212·5 mM/l there was a rapid shrinkage of the cells to 55 per cent of their original volume. This was accompanied by an equally rapid rise in $[Na^+]_i$ to six times the initial value with a further rise to 13·7 times after 160 minutes. Similar experiments with potassium resulted in a sharp increase of $[K^+]_i$ to 1·75 times, and subsequently to 4·5 times, the initial value. There were no changes in $[K^+]_i$ when $[Na^+]_o$ was raised nor in $[Na^+]_i$ when $[K^+]_o$ was raised. Curiously, increase in the external concentration of sodium or potassium chloride did not lead to an increased internal chloride concentration. Careful examination of these and other results convinced Dunham and Child that sodium and potassium in *Tetrahymena* are separated into compartments. That is to say, not all the sodium or potassium in the animal is equally free to diffuse either within the animal or across the pellicle. They calculate that there is one compartment of $[Na^+]_i$ of 1·9 m eq/l which is inexchangeable with the medium. A second compartment of 3·1 m eq/l is rapidly exchangeable but does not vary in size with changes in $[Na^+]_o$. The third compartment is freely exchangeable but changes proportionally if $[Na^+]_o$ is changed. There was also evidence for an active mechanism for sodium extrusion. Some of this extrusion probably takes place through the contractile vacuole. Dunham (1969) showed that if $[Na^+]_o$ was dropped from 20 to 2 mM, $[Na^+]_i$ decreased from 10 to 4 mM/kg of cells in 3 minutes. However, this fall in $[Na^+]_i$ was blocked if the contractile vacuole was inhibited through the addition of 100 mM sucrose at the same time as the $[Na^+]_o$ reduction. With $[Na^+]_o$ constant inhibition of vacuolar activity resulted in a rise in $[Na^+]_i$ from 10 to 27 mM/kg of cells, suggesting that the contractile vacuole is at least one of the means of sodium extrusion. This aspect of contractile vacuole physiology deserves further attention. The need for sodium extrusion may explain the presence of contractile vacuoles in some marine ciliates. For instance Kehlenbeck *et al.* (1965) reported that a marine species of *Uronema* had a contractile vacuole whose rate of discharge fell with increase in external salinity but which was still active when the animal was in 300 per cent sea-water. Even under these conditions the intracellular salt concentrations were not high and, although aminoacid concentration was not measured, it is unlikely that the function of the vacuole was only osmoregulation. A role in ionic regulation is much more probable.

Results similar to those for *Tetrahymena* have been obtained from

Spirostomum (Carter, 1957). He studied the movements of radio-active sodium, potassium and bromine into and out of this ciliate. He found a relatively constant $[K^+]_i$ despite changes in $[K^+]_o$. With a $[K^+]_o$ of $0 \cdot 65$ mM/l the $[K^+]_i$ was $7 \cdot 0$ mM/l of cells. *Blepharisma* (Hilden, 1970) has a similar distribution of sodium and potassium. $[Na^+]_i$ is maintained at a low level and is relatively unaffected by rises in $[Na^+]_o$ while $[K^+]_i$ is held at a constant level above $[K^+]_o$.

Investigations with the marine ciliate *Miamiensis aridus* and *Uronema filificum* (Kaneshiro *et al.,* 1969a,b) have shown that these species have internal salt concentrations not very dissimilar to fresh-water forms. In 100 per cent sea-water *M. aridus* contains $87 \cdot 9$ mM/kg cells of sodium, $73 \cdot 7$ mM/kg cells of potassium and $60 \cdot 8$ mM/kg cells of chloride. Dilution of the external medium of 50 per cent sea-water resulted in these values falling to $52 \cdot 8$, $60 \cdot 6$ and $24 \cdot 9$ mM/kg cells respectively. Very similar figures were obtained for *U. filificum*. *M. aridus* maintains itself hyperosmotic over a wide range of sea-water concentrations. This is accomplished by changes in internal sodium and potassium concentrations and also by appropriate changes in the free aminoacid levels (see p. 120).

Although there is a considerable mount of information on sodium, potassium and chloride concentration in ciliates other ions have been relatively neglected. Calcium distribution has been studied in *Spirostomum ambiguum*. This heterotrich forms granules of apatite (calcium phosphate) (Pautard, 1959; 1970). Work with radioactive calcium has shown that $[Ca^{++}]_i$ may be as high as $14 \cdot 6$ mM/l cells as compared with a $[Ca^{++}]_o$ of only $0 \cdot 5$ mM/l (Jones, 1966). However, only 10 per cent of this calcium is readily exchangeable. It is probable that there are three compartments for calcium in this animal and that in two the calcium is bound (Jones, 1967). The accumulation of radioactive phosphorus parallels that of calcium and presumably represents calcium phosphate formation. About 10 per cent of the phosphorus is exchangeable but there seem only to be two compartments. The uptake and loss of ^{45}Ca by *Paramecium* (Yamaguchi, 1960) suggest that in this animal calcium is freely exchangeable, and that calcium competes with potassium for either a means of entry into the cell or for binding sites within it. $[Ca^{++}]_i$ is maintained slightly above $[Ca^{++}]_o$. In *Miamiensis avidus*, $[Ca^{++}]_i$ and $[Mg^{++}]_i$ are held more or less constant in the faces of large changes in external concentration. $[Ca^{++}]_i$ is about $5 \cdot 0$ mM/kg cells and $[Mg^{++}]_i$ 30 mM/kg cells (Kaneshiro *et al.,* 1969a).

A good general picture of cationic regulation in ciliates emerges from these studies. Potassium is held at a level above that of the environment in the case of freshwater ciliates. This accumulation may be due either to a Donnan equilibrium, binding to the cellular

I20   The Ciliates

anions, transportation across the membrane reciprocally with sodium or an active accumulation of potassium. There are insufficient data on the relation between membrane potential and $[K^+]_i$ to evaluate the role of a Donnan equilibrium, but it must play some part. It is equally likely that some of the potassium is bound inside the cell, although Riddle (1962) had evidence that all of the internal potassium of the giant amoeba, *Pelomyxa carolinensis,* is ionized. In contrast Klein (1961) reported that in the smaller *Acanthamoeba* only 40 per cent of the potassium was unbound. In *Tetrahymena* some potassium seems to be bound but Carter obtained no such evidence from *Spirostomum*. Indeed his results for both potassium and cloride indicate a Donnan distribution of these ions. Results from *Blepharisma* (Hilden, 1970) indicate that metabolic energy and a reciprocal movement with sodium may play some part in maintaining a high $[K^+]_i$. Thus a variety of means seem to be employed by protozoa in keeping up their potassium content. Sodium, on the other hand, is maintained at a low level in the cell and there is evidence that it is actively extruded, and, in one case at least, the contractile vacuole appears to play a leading role in the extrusion. Andrus and Giese (1963) found that at low temperatures *Tetrahymena* gained sodium from and lost potassium to the environment. The normal distributions of these ions were restored when the animals were returned to room temperature. Sodium extrusion seemed to be insensitive to many metabolic poisons, e.g. iodoacetic acid, sodium azide and dinitrophenol. Even oubain, a specific inhibitor of the sodium pumps of other cells, failed to have any effect. All of the drugs except oubain resulted in a loss of potassium from the cell. Andrus and Giese suggest that there is active control of intracellular sodium and potassium levels in *Tetrahymena* and at least some of the ion movements may be via a coupled mechanism moving potassium inwards and sodium outwards. In both *Tetrahymena* and *Spirostomum* part of the sodium content is relatively inexchangeable.

There are fewer data on anions. Bromine in *Spirostomum* conforms to a Donnan distribution but in *Tetrahymena* chlorine is held steady over a wide range of $[Cl^-]_o$.

INTERNAL CONCENTRATION OF AMINOACIDS

Until recently aminoacids have been largely ignored in assumptions about internal osmotic pressure of ciliates. However, it is now known that they make an important contribution in some cases. In the marine ciliate *Miamiensis aridus* the total free aminoacid content is 317 mM/kg cells when the animal is in 100 per cent sea-water (Kaneshiro *et al.,* 1969b). The most important aminoacids are alanine, glycine and proline which account for 73 per cent of the total. A reduction of the external concentration to 25 per cent sea-water results in a reduction of the free aminoacid concentration in the ciliate to 24 per cent of that when in 100 per cent sea-water. Animals in 200 per cent sea-water have 22 per cent more amino-acids than those in 100 per cent sea-water. The changes in amino-acid levels result from the mobilization of bound molecules and are complete in 70 minutes. *Miamiensis* maintains itself hypertonic to the environment over a wide range of salinities and it is suggested that this ensures a supply of osmotic water resulting in continuous contractile vacuole activity. Such activity could well be of signi-ficance in regulating internal sodium concentration (see p. 118 and Dunham, 1969.)

In *Tetrahymena,* also, aminoacids are apparently mobilized to ensure that the animal is hypertonic to its medium (Stoner and Dunham, 1970). An addition of 100 mM/l sucrose to the medium results in a rise in the aminoacid concentration from 56·5 to 114·9 mM/kg cells. (This is an approximation, as there is a degree of cell shrinkage after the addition of sucrose. The figures in fact represent concentration per $8·52 \times 10^{10}$ cells.) The rises in internal salt con-centration are much smaller although sodium does increase, pre-sumably as its expulsion via the contractile vacuole temporarily ceases. These recent studies emphasize the importance of taking aminoacids into account when calculating internal osmotic pressure of ciliates.

MECHANISM OF SYSTOLE

For the contents of the contractile vacuole to be expelled to the exterior, the hydrostatic pressure inside the vacuole must exceed that outside the animal at systole. This pressure difference could arise in two ways; either by tension developed in the vacuole wall or from turgor in the cytoplasm around the vacuole. The former could result from an inherent tension in the membrane itself or from contractile structures in or near the vacuolar wall.

As Kitching (1956a) has pointed out, the vacuoles of *Paramecium* continue to function even when the animal is flattened by exosmosis, as do those of *Podophrya* whose outer surfaces are wrinkled (as happens just after food capture). These facts argue against the necessity of turgor for effective systole. However he also notes that *Podophrya* grown in dilute sea-water and transferred to freshwater exhibit very brisk systole which can presumably be explained by a high turgor pressure.

Wigg et al. (1967) claimed that the contractile vacuole of *Amoeba* was flattened during systole by the turgor of the cytoplasm. This claim was supported by similar evidence from *Paramecium* (Organ *et al.*, 1968b). In the latter case the authors claim that the vacuole is indented while emptying, a configuration that could only result from collapse under pressure. They suggest that the contractile vacuole should be renamed the 'water expulsion vesicle'. Although their pictorial evidence is sometimes unclear it is unlikely that anyone would deny that body turgor *contributes* to systole. However such evidence does not refute elastic or contractile properties of the vacuolar wall. Film analysis of systole in animals lacking turgor due to exosmosis would be instructive.

Two interesting pieces of work from Dunham's laboratory tend to oppose the work of Organ *et al.* (1968b). Dunham and Stoner (1969) showed that the contractile vacuole of *Tetrahymena* becomes spherical just before systole and apparently remains so during systole. This they claim 'may be inconsistent with the proposal that . . . the force is generated in the adjacent cytoplasm and not by the wall of the vacuole'. Prusch and Dunham (1970) were able to isolate contractile vacuoles from *Amoeba*. These vacuoles appeared to be free of adhering cytoplasm. The application of solutions containing ATP and magnesium chloride caused the vacuoles to contract. In the presence of calcium the induced contraction is irreversible. At least for this animal the contractile vacuole seems to be intrinsically contractile and to be stimulated by a combination of chemical agents which are active in other contractile systems. It is interesting to note that Organ *et al.* (1968a) reported an increase in activity of the contractile vacuole of *Paramecium* when ATP was applied externally to the animal.

Although more work is required in this field it would appear at present that both turgor and contraction play some part in systole, the former perhaps being an important contributing force not always required.

Under steady conditions, the frequency of systole and the ultimate size of the vacuole remain fairly constant. Kitching (1956a) has pointed out that this offers two alternative explanations as to the timing of systole. Either there is a rhythm of activity determining the frequency of systole or there is a critical size at which the vacuole discharges. However, if the osmotic conditions are changed both the frequency and the ultimate size vary. It may be that both factors play some part in determining the time of systole. A critical factor at the onset of systole would seem to be the unplugging of the pore or canal through which the vacuole discharges. This may be effected by fibres found around such pores (Elliott and Bak, 1964a; Rudzinska, 1958). Organ *et al.* (1968b) report that in *Paramecium* the plug is 'ripped along one semicircular border and is driven against the opposite wall as a flap while the water rushes out'. This suggests that the pressure inside the vacuole forces open the pore. However, it is more likely that the pore is opened by some active process. Dunham and Stoner (1969) found that just prior to systole in *Tetrahymena* the pellicle at the vacuole pore became indented. They suggest that the indentation may reflect a shortening of the fibrils which are arranged around the discharge canal. These fibrils may well be responsible for rupturing the pore membrane. In the case of one particular animal that they observed, systole ceased, for unknown reasons, for about a minute. Dunham and Storer observed that during this time periodic indentations of the pellicle at the site of the pore continued. Eventually normal systoles were resumed. This suggests a rhythmic activity associated with pore opening and supports, to some extent, the first of Kitching's two suggested control mechanisms.

DIASTOLE AND FINE STRUCTURE

Very little is known as to how the fluid is excreted into the vacuole. Various thoughts on this matter have been reviewed by Kitching (1956a). The fine structure of ciliate contractile vacuoles and collecting ducts indicates that fine tubules, continuous with the endoplasmic reticulum of the animal, may play some part in the production of vacuolar fluid. In some ciliates, e.g. *Zoothamnium* (Carasso *et al.*, 1962) these tubules apparently connect with the vacuole itself and are of two different sizes. They form a mass of tubular material around the vacuole (the osmophilic layer of the light microscopists) termed the spongiome. In other ciliates these tubules are continuous not with the vacuole itself but with collecting ducts. This has been well documented in the case of *Paramecium* by Schneider (1960) (Fig. 28). In this case it would seem that the vacuolar fluid passes from tubules, continuous with the endoplasmic reticulum, into the collecting ducts, each of which has a spongiome-like region around it. The collecting ducts become distended and by some means expel their contents into the vacuole itself which becomes distended. Thus the systole of the ducts causes the diastole of the vacuole. The connection between the fine tubules and the ducts is severed during duct systole presumably preventing back-flow of fluid into the tubules. A similar function is probably performed by the constriction that appears at the junction of the ducts and the vacuole during the latter's systole. There are various fibre systems connected with the vacuole, its collecting ducts and outlet canal. It has been suggested that fibres in the wall of the vacuole are contractile but they may well have a supporting role, preventing damage to the junctions with the

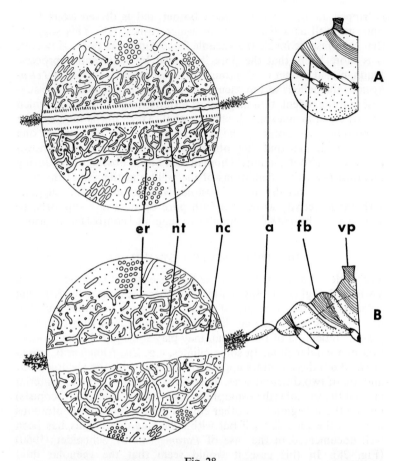

Fig. 28
Structure of the contractile vacuole and one nephridial canal of
Paramecium aurelia. **A,** nephidial canal in systole, contractile vacuole
in diastole. **B,** nephridial canal in diastole, contractile vacuole in
systole. a, ampulla; er, endoplasmic reticulum; fb, fibrillar elements;
nc, nephridial canal; nt, nephridial tubule; vp, vacuole pore opening
to the exterior. After Schneider (1960).

collecting ducts at the time of vacuole systole. The photographs taken
during systole by Organ *et al.* (1968b) suggest that the half of the
vacuole around the outlet canal retains its shape during systole. This
may be because of the fibrils in the vacuole wall. The fibres around
the outlet canal and vacuolar pore are more likely to be contractile
and probably play some part in opening the pore.

EXCRETION OF NITROGENOUS WASTES

Early workers were tempted, by analogy with the vertebrate kidney, to assume that the contractile vacuole was responsible for the excretion of dissolved nitrogenous wastes. Various claims to this effect have been made but it now seems clear that many ciliates excrete ammonia which probably diffuses rapidly out through the pellicle (Kidder, 1967). Kitching (1967) admits that some nitrogenous wastes might be excreted via the contractile vacuole and that this possibly explains the possession of vacuoles by marine protozoa. However, in view of the work (reviewed p. 118) from Dunham's laboratory it is most likely that the vacuoles of marine ciliates are more involved with sodium regulation than nitrogenous excretion.

6

REPRODUCTION AND GROWTH: I

ASEXUAL REPRODUCTION

Ciliates reproduce in two ways, asexual and sexual. Asexual reproduction occurs in a variety of ways.

Binary fission to produce similar daughter cells
In this case the ciliate divides into two parts, each of which looks like a small, but essentially complete, adult (Plate 4). These grow and in time divide again. The plane of fission in ciliates has often been termed transverse; that is to say across the animal's long axis, in contradistinction to the longitudinal division of flagellates. In ciliates of simple body form this description holds good, but when the morphology is more complicated it is unsatisfactory. In peritrichs, for instance, division superficially appears to be longitudinal. Corliss (1961) prefers the term homothetogenic to describe ciliate division. This implies that while the daughter cells may be identical they are not mirror images of one another as is the case in flagellates (see Fig. 29). In those cases where division obviously cuts across the kineties the term perkinetal may be used to describe fission. The anterior daughter cell is called the proter and the posterior one the opisthe.

Some ciliates reorganize their structure prior to binary fission. For example *Uronychia,* a hypotrich, resorbs its cirri; there is a dedifferentiation of the peristome region in *Bursaria,* while in *Chilodonella* the pharyngeal basket, cytostome and body cilia are resorbed. In some apostomes there may be considerable changes in the arrangement of the somatic ciliature. In *Gymnodiniodes* the kineties of the trophont undergo changes that result in a much simpler pattern at the time of fission (Fig. 34) (Lwoff, 1950). However, in

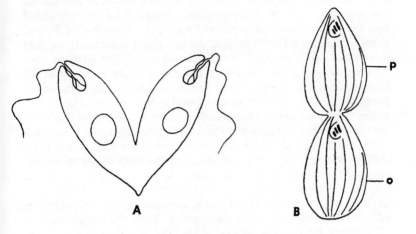

Fig. 29
Diagram to show typical mode of division in flagellates (**A**) and ciliates
(**B**). p, protor; o, opisthe.

many ciliates the organelles remain intact. In either case, before or
shortly after division the adult complement of organelles must be
doubled. The nuclei must divide, the number of somatic cilia must
double and two complete oral apparatuses must be provided for the
daughters.

Micronuclear division appears to be by mitosis (Grell, 1964)
although clear description is difficult due to the small size and almost
spherical shape of the chromosomes. Macronuclear division is
amitotic. The macronucleus elongates and constricts into two roughly
equal parts. This may be preceded by a simplification of macronuclear structure. In *Spirostomum ambiguum* and *Stentor coeruleus* where
the macronucleus is moniliform, it first condenses to an ovoid, then
divides. The C-shaped macronucleus of *Euplotes* contracts in a
similar way.

The macronucleus is usually large (Fig. 6), occupying 50 or more
times the volume of the micronuleci in some cases, and contains
large amounts of DNA. It is obviously a multiple nucleus. As the
nucleus divides by a simple constriction without any apparent
mitotic figures, the problem arises as to how each daughter receives
a complete genome (set of genetic instructions) (Nanney and Rudzinska, 1960). It is apparent from regeneration studies that even small
pieces of macronucleus have the capacity to form a complete and

fully functional macronucleus. This implies that any small fragment must contain at least a single genome and that probably a complete macronucleus contains many genomes. Sonneborn (1947) suggested that the macronucleus is organized into a large number of diploid subnuclei. Nilsson (1970) produced some visual evidence to support this theory. He examined the macronucleus of *Tetrahymena pyriformis* strain GL (which is amicronucleate) at various stages in the life-cycle. During interphase the nucleus contained many chromatin granules. Before division these granules become associated into chains which in turn condense and spiral into large aggregations of chromatin. Nilsson claims that these aggregates are haploid genomes. There are usually about 80 such aggregates in a nucleus. Falk *et al.* (1968) have suggested that the microtubules seen in the macronucleus of *Tetrahymena* may serve to attach individual genomes to the nuclear envelope ensuring that both daughters receive almost equal numbers of genomes. These microtubules may have an alternative or additional skeletal or contractile function. There is a certain amount of evidence from genetical and other studies (Allen and Nanney, 1958) to support the subnuclei theory.

The increase in number of somatic cilia through division has been studied in *Tetrahymena* by William and Scherbaum (1959). They found that the numbers of kinetosomes in a particular kinety were 29–60 in a normal population of animals. In animals about to divide the mean number of kinetosomes was 58. During division it was impossible to count the kinetosomes accurately. The mean number just after division was 34. This suggests that during division the kinetosomes increase from 58 to 68 (2×34) indicating that the proliferation of somatic cilia is particularly rapid during the relatively short cleavage period. This period of the cell cycle is also marked by an increase in respiration. In non-dividing cells the authors noted that many kineties contained kinetosomes without, or with only a short cilium and interpreted these as recently formed basal bodies.

In many ciliates the anteriorly placed oral apparatus is inherited by the proter (e.g. *Tetrahymena*). This means that there must be a new set for the opisthe. As with other organelles it is of obvious selective advantage to have the new system functional at or soon after division. It is not surprising, therefore, that stomatogenesis often occurs before cleavage. Tartar (1961, paraphrasing Johnson, 1893) has written that 'the situation in ciliates is the reverse of that in metozoa since all the "embryology" occurs before instead of after reproduction'. In *Tetrahymena* stomatogenesis starts with the proliferation of kinetosomes in a small patch about half-way down kinety 1. At first this anarchic field is simply a crowded patch of kinetosomes. Gradually, however, they become organized into the bases of three

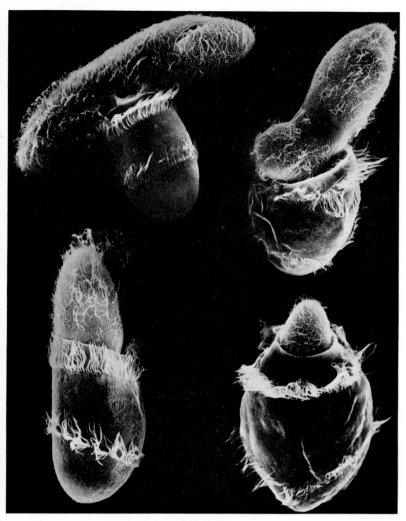

PLATE 3

Series of scanning electron micrographs showing the capture and ingestion of *Paramecium* by *Didinium*. The top left picture shows the animals shortly after collision. The proboscis of the *Didinium* is beginning to dilate. In the top right the *Paramecium* is being bent as it is drawn into the proboscis. A later stage is shown in the bottom left. In the bottom right, the prey is almost completely engulfed by the now much distended predator. Photographs kindly supplied by Drs. H. Wessenberg and G. Antipa, and reproduced by courtesy of the *Journal of Protozoology*. Magnification about × 350.

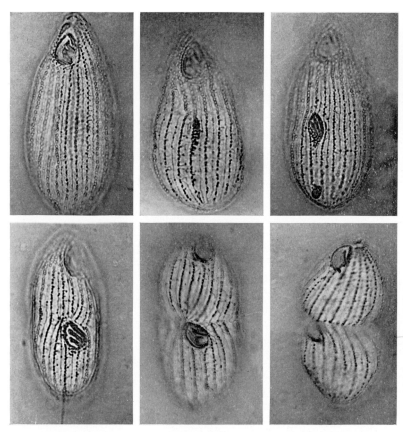

PLATE 4

Division and stomatogenesis in *Tetrahymena pyriformis*. This series of photographs shows the gradual development of the typical tetrahymenid oral ciliature by proliferation of kinetosomes from the stomatogenic kinety. The pictures should be read from top left (interdivision adult) to bottom right (almost completed division). Primary and secondary meridia can be seen (see p. 42). Silver stained.

membranelles and the undulating membrane. At first there is no depression at the site of the developing oral apparatus and it is only at the time of division that the buccal cavity develops.

A similar but somewhat more complicated series of events can be seen during binary fission in *Stentor coeruleus* (see Fig. 30). An anarchic field develops about half-way down the animal below the oral pouch and gradually grows to form a crescent bearing short membranelles. An area free of pigment granules, the future oral pouch region, appears at the posterior end of the newly formed AZM. At the anterior of the AZM severing of pigment stripes and kineties is initiated and progresses round the animal marking the future cleavage furrow. The anterior end of the new AZM also migrates posteriorly disrupting the pigment stripes, so producing the characteristic pattern that will distinguish the opisthe from the proter for some time after division. The stentor then constricts along the cleavage furrow and finally separates into two. During division there is also a certain amount of resorption and reorganization of the proter's AZM.

Binary fission to produce dissimilar daughter cells

In some cases division may result in one of the daughter cells being different from a typical adult. This is commonly, but not always or only, the case in sessile protozoa. In Suctoria for instance various types of ciliated buds are produced by the non-ciliated adults. The bud may be formed internally in a brood pouch which results from an invagination of the outer surface (e.g. *Tokophrya,* Millecchia and Rudzinska, 1968). In other genera the larvae are produced externally; a portion of the adult becomes ciliated and is then nipped off. In some cases many larvae are produced at the same time. Such multiple budding may be either internal (e.g. *Trichophrya*) or external (*Ephelota*). Other groups of sessile ciliates also produce a free-swimming distributive larva. This is the case with stalked peritrichs. The larva is called a telotroch and has, as well as the oral cilia, a posterior ring of locomotory cilia, the trochal band. In the course of evolution, neoteny has produced forms which complete their life-cycle in this free-swimming stage and for which sessile forms are unknown.

The preconjugation division in peritrichs also results in two dissimilar daughter cells. The smaller microconjugant swims off to seek out the still sessile macroconjugant. Unequal division is also found in some normally free-swimming forms. This is true in the rumen ciliate *Opisthotrichum janus* where the larger daughter inherits the tough skeletal plate (which presumably prevents a more equal division). The astome *Haptophrya* divides transversely to produce

E

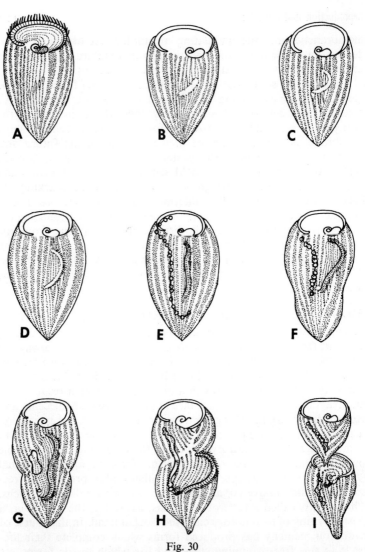

Fig. 30

Division in *Stentor coeruleus*. **A,** granular bands split; first indication
of division. **B,** oral primordium appears as a lateral rift in ectoplasm.
C, primordium elongates anteriorly. **D,** first appearance of new membranelles. **E,** primordium now full length, new membranelles beat in
slow metachronal rhythm. **F,** site of future buccal cavity apparent;
macronucleus begins to condense. **G,** posterior end of new AZM coils
inward sharply; stripes severed around cell; macronucleus fully con-
densed. **H,** new AZM migrates posteriorly leaving herring-bone 'scar'
in proter; macronucleus begins to divide. **I,** division almost complete,
opisthe mouthparts almost completely functional; macronucleus
becomes nodulated. After Tartar (1961).

one large and one small daughter. However, the offspring do not
separate. Subsequent growth and division produce a chain of such
buds (see Fig. 31). In *Polyspira delagei* the daughters are identical
and a chain of 16 buds may be produced.

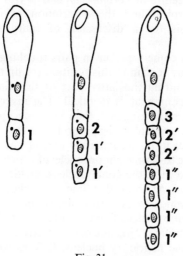

Fig. 31
Budding in *Haptophrya*. After Dogiel (1965).

Division in cysts

This is a frequent occurrence. There is usually more than one
division in such cysts. The fish parasite *Ichthyophthirius multifilius*
divides repeatedly inside a cyst and as many as 2000 offspring may
emerge. The common trichostome *Colpoda* divides in a thin-walled
cyst to produce usually 4, sometimes 2, 8 or 16 daughters.

SEXUAL REPRODUCTION

Sexual reproduction in metazoa involves the fusion of haploid
gametes, usually differentiated into egg and sperm. The fusion of
their nuclei produces a diploid synkaryon containing the chromosomes of both gametes. Sexual reproduction in ciliates, conjugation,
differs from that of metazoan in that there are no gametes.* Two

*It could perhaps be argued that the animals themselves act as gametes. This is
certainly the case with some unicellular algae where the adults come together in
pairs, fuse and form a zygote. In the case of ciliates, however, the individuals do
not lose their identity (except in a few cases where the conjugants are specialized,
e.g. peritrichs) and right up to the time of nuclear exchange contain both 'male'
and 'female' nuclei. After nuclear fusion it would be reasonable to regard each
of the ciliates as a zygote.

adults come together and reciprocally exchange haploid nuclei. There is usually no sexual dimorphism (though see account of mating types below) but in some instances (e.g. sessile peritrichs) the conjugants may differ somewhat. Although there are many specific variations, conjugation in *Paramecium aurelia* (for review see Beale, 1954) is fairly typical and the description of it which follows will form a basis for a discussion of more specialized types (Fig. 32).

The paramecia come together in pairs touching along their oral surfaces. They may remain in this position for up to 15 hours. Conjugation occurs during the latter part of this time, and the part played by each conjugant is identical. The macronucleus disintegrates, first becoming elongated, then breaking up into many small fragments. These gradually disappear but a few may persist after the end of conjugation. *Paramecium aurelia* has two micronuclei. These divide twice to form eight pronuclei of which seven degenerate. The survivor, known as the gonal nucleus, divides again to produce two gametic nuclei. At about this time a cytoplasmic bulge forms in the oral region which fuses with a similar bulge in the other conjugant. One gametic nucleus from each conjugant migrates through this cytoplasmic bridge. This nucleus is referred to as the 'male' or, more correctly, the migratory nucleus. It fuses with the stationary (or 'female') nucleus of its partner. The conjugants now separate and the diploid synkaryon divides twice so that each ex-conjugant has four nuclei; two become micro- and two macronuclei. At the first cell division after conjugation, the micronuclei divide but not the macronuclei and each daughter inherits two micronuclei and one macronucleus, the correct complement. Subsequent fission of course involves division of all nuclei.

There has been considerable discussion as to which of the three nuclear divisions leading to the production of the gametic nuclei are reductional (Sonneborn, 1947). Genetical evidence favours the first two being reductional and the last being equational. Thus the gametic nuclei are sisters (and have identical genotypes).

There is considerable variation in conjugation even within a single genus. In *Paramecium caudatum*, for example, there is only a single micronucleus, and thus only four nuclei are produced by reductional divisions, of which three degenerate to leave the gonal nucleus. Fusion of the gametic nuclei is as in *P. aurelia* but in the exconjugant there are three nuclear divisions to yield eight nuclei. Of these three break down to leave five of which four become macronuclei and one a micronucleus. The exconjugant now divides twice with accompanying division of the micronucleus but not the macronuclei. The four cells which result each have one macronucleus and one micronucleus,

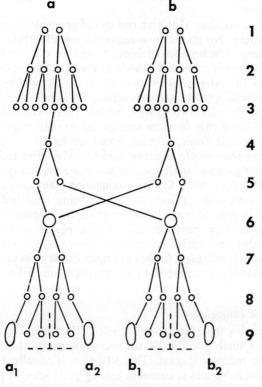

Fig. 32

Conjugation in *Paramecium aurelia*. Two animals shown in position prior to conjugation. a and b show the sequence of events for their respective micronuclei. Each conjugant has two micronuclei (1) which divide twice (2 and 3) to yield 8 pronuclei of which 7 degenerate. The gonal nucleus (4) divides to form 2 gametic nuclei (5). One of these migrates into the other conjugant and fuses with its stationary gametic nucleus to form a zygotic nucleus (6). The conjugants separate, and the synkaron divides twice (7 and 8). There is now a cell division and a further nuclear division to yield offspring with one macro- and two micronuclei. The two conjugants (a and b) have produced 4 offspring ($a_1a_2b_1b_2$).

the normal state. Of course between genera variations in conjugation are even greater (see Dogiel, 1965, for review).

Macronuclear disintegration in the conjugant may occur at the beginning of conjugation (e.g. *Bursaria truncatella*), in the latter half of conjugation (e.g. *Paramecium aurelia*) or postponed until the end (e.g. *Tetrahymena pyriformis*). In the suctorian *Collinia branchiarum* there is an exchange of macronuclear fragments during conjugation. In *Paramecium* kept under experimental conditions it is possible for fragments of macronucleus to survive conjugation when they regenerate.

The meiotic divisions that give rise to the pronuclei have a fairly constant pattern. No matter how many micronuclei there are, they usually undergo synchronous division. In species with many micronuclei (e.g. *Bursaria truncatella*) this may result in the production of nearly 50 nuclei. After meiosis these nuclei usually undergo a third division after which all but one degenerate. This single gonal nucleus divides to produce the two gametic nuclei. There is cytological and genetical evidence that in some species the gametic nuclei are not sisters but derived from different gonal nuclei (Grell, 1967). In *Trachelocerca phoenicopterus* many nuclei survive after meiosis and produce many gametic nuclei, some of which are migratory and some stationary (Raikov, 1958). During conjugation many synkaryons are formed but only one survives, the others being absorbed. Raikov interprets this type of conjugation as primitive, representing a recapitulation of true gametogenesis. Post-conjugation nuclear behaviour is also very varied as has been indicated for *Paramecium* species. Dogiel (1965) gives further examples of the ways in which the nuclear apparatus is restored to its normal state following conjugation.

Anisogametic conjugation

In some species there are morphological differences between the conjugants. Usually one is small, the microconjugant, while the other is large, the macroconjugant. This situation is usually associated with a sessile habit and is common among peritrichs. In this latter group the macroconjugant has the appearance of a normal adult complete with stalk while the microconjugant is small and motile. The karyology of their conjugation is complex and varied (Grell, 1967). In some cases there is full exchange of gametic nuclei but no nuclear fusion in the microconjugant. In others the stationary nucleus of the microconjugant and the migratory nucleus of the macroconjugant degenerate. In yet others there is no division of the gonal nuclei. That of the microconjugant migrates into its partner where the two fuse to form a synkaryon. In all cases the microconjugant fails to form a viable exconjugant and is absorbed by the sessile partner. In Suctoria conjugation may take place between two stalked individuals but one may become detached from its stalk and absorbed into the other (e.g. *Ephelota gemmipora*).

PSEUDOSEXUAL BEHAVIOUR

Two forms of nuclear behaviour related to conjugation sometimes occurs in ciliates. In *cytogamy* the animals may come together in pairs and go through all the actions that lead to the production of gametic nuclei. At this stage, however, instead of a reciprocal exchange the sister gametic nuclei of each individual fuse to form a synkaryon. Subsequent reconstitution of the normal nuclear apparatus is the same as in the exconjugant. During *autogamy* similar nuclear behaviour is observed in isolated individuals. As in both autogamy and cytogamy the gametic nuclei are derived by mitosis from the gonal nucleus it is often assumed that the resulting synkaryon is homozygous for all genes. However, this is not always the case. Heterozygous post-autogamous animals have been detected in some strains of *Euplotes minuta* where autogamy probably takes a slightly different course to that in *Paramecium aurelia*.

MATING TYPES

Sonneborn (1937) discovered that conjugation of *Paramecium aurelia* only occurs between animals of different and compatible mating types. This species is now known to be divided into at least 16 varieties, or syngens, most of which contain two mating types. Conjugation will only take place (except under exceptional conditions, Sonneborn, 1947) between individuals that belong to the same syngen but different mating types. In *P. bursaria* 6 syngens are known, 1 with 8 mating types, 3 with 4, 1 with 2 and 1 in which conjugation is unknown. Similar situations exist in other ciliates and mating types have been demonstrated in *Tetrahymena, Colpidium, Tokophrya, Euplotes, Oxytricha* and *Stylonichia* (Preer, 1969).

It has been suggested that mating types constitute a 'self-sterility' system like that found in higher plants, but autogamy shows that ciliates are self-fertile. Mating types have also been equated with sexes, but Beale (1954) points out that this analogy also is misleading as ciliates produce both a stationary ('female') and migratory ('male') gametic nucleus. Each individual is therefore hermaphrodite. Beale is of the opinion that it is 'probably best to consider the mating types of *Paramecium* as peculiar to the genus (or at least to the group of ciliates), and avoid attempts at making analogies with other groups'.

The discovery of mating types permitted studies of their inheritance (see Beale, 1954, and Preer, 1969, for reviews). Most of the work has been done with *Paramecium aurelia* where there are two ways in which mating types are determined. In group A (including syngens 1, 3, 5, 9 and 11) the inheritance is termed caryonidal. Each exconjugant contains a synkaryon which divides to form 4 nuclei of which 2 are presumptive macronuclei. At this point cell division takes place and each daughter receives one of these macronuclei. Each of these animals is called a caryonide and its descendants will receive its macronuclear material by division. Thus conjugation produces 4 caryonides (2 from each conjugant). The organisms which descend from a caryonide are all of the same mating types but the 4 lines resulting from a conjugation may belong to different mating types. The proportion of mating types produced is no different from that expected by chance nor is it influenced by the mating types of the cytoplasmic parent. Thus it seems that mating types in this group are determined by macronuclear factors. Something (as yet unknown, although Preer (1969) offers some speculations) happens at the time of the first cell division after conjugation to 'fix' the caryonidal macronucleus so that cells containing it belong to a particular mating type. Whatever produces the macronuclear variation can be influenced by a change in temperature shortly after conjugation (i.e. at the time of formation of the new macronuclei) (Sonneborn, 1939).

In group B (which includes syngens 2, 4, 6, 7, 8, 10 and 12) the offspring have the mating type of their cytoplasmic parent, although there is evidence that caryonidal inheritance plays some part (Preer, 1969). It has been established that control of mating type in group B originates from the macronucleus which is itself influenced by the cytoplasm.

7

REPRODUCTION AND GROWTH: II

THE CELL CYCLE. DNA SYNTHESIS

For the greater part of a ciliate's existence, feeding and growth lead to binary fission followed by more feeding and growth. Sexual reproduction is a relatively infrequent occurrence. During the period between divisions the animal must double many of its constituents. Most importantly it must double the amount of DNA in its macro- and micronuclei. The cell cycle is conventionally divided into various phases. The first phase is that between the completion of division and the onset of DNA synthesis. This is called G_1. The period DNA synthesis is called S and the time from the end of DNA synthesis to division is termed G_2. D is the time of division. Among cell types there is marked variation in the relative length of these phases, but they usually are of constant length for any given cell type. Investigations with tritiated thymidine* have shown that in the macronucleus of *Tetrahymena pyriformis* strain HSM (Prescott and Stone, 1967) G_1 takes up 30 per cent and S 35 per cent of the cell cycle. (The mean generation time in these experiments was 3·7 hours.) Division takes about 10 per cent of the cycle. The micronucleus behaves quite differently. The synthesis of DNA takes place during the late telophase of its mitosis, that is during the time of cytokinesis, and continues for a short time after the separation of the daughter cells. Micronuclear S therefore does not overlap with macronuclear S.

---

*Thymidine is a nitrogenous base required for the synthesis of DNA (but not RNA). Radioactive tritium (^{3}H) can be substituted for hydrogen (^{1}H) in this molecule and the presence of labelled base detected by autoradiography. Incorporation of the labelled molecules into nuclei is taken to indicate DNA synthesis. The presentation of tritiated thymidine to the cell at various times during the cell cycle can be used to determine the onset and duration of S.

If the cultural conditions, especially temperature and nutrients, are altered the time of the various phases, as well as the total time for the cycle, are altered. Considering the macronucleus, McDonald (1962) found that if the generation time is lengthened the increase appears in G_1, while S and G_2 remain unchanged. However, if *Tetrahymena* in S are transferred from a nutrient medium into one lacking proteose peptone the period of DNA synthesis is lengthened by 80–90 minutes as compared with control cells (Prescott and Stone, 1967). If the cells are transferred to a solution lacking two essential aminoacids S can be lengthened by up to 200 minutes. Cell division, however, is not delayed so that S almost merges into D.

In *Paramecium aurelia* the times of DNA synthesis are very different from *Tetrahymena* (Kimball and Perdue, 1962; Kimball *et al.,* 1960). G_1 for both nuclei is about 50 per cent of the cell-cycle. Micronuclear S takes only 9 per cent of the interdivisional time but macronuclear S continues almost up to the beginning of D. In *Euplotes eurystomus* micronuclear S is similar to that in *Tetrahymena*, lasting from telophase to just after nuclear division. As has been described on p. 59, macronuclear DNA synthesis starts at either end of the long C-shaped nucleus and progresses along the nucleus until the two 'reorganization bands' meet in the middle and synthesis ceases (Gall, 1959; Prescott *et al.,* 1962). This macronuclear S lasts for about 75 per cent of the cell cycle and there is no G_2. *Euplotes* which have been starved cease reproducing and come to rest in G_1. In *Stentor coeruleus* macronuclear S lasts some 90 per cent of the cell cycle during rapid growth (Guttes and Guttes, 1960).

Much effort has gone into trying to discover the factors which control the onset of division and initiation of DNA synthesis. As the beginning of DNA production is the earliest detectable preparation for division the transition from G_1 to S has been closely studied. Work with ciliates has contributed significantly to this field. For example, if *Tetrahymena* is deprived of the aminoacids tryptophan and histidine during G_1, DNA synthesis is reduced to 20 per cent of its normal amount and the animal never divides. If, however, the aminoacids are withdrawn after the beginning of S, the DNA synthesis continues as normal and cell division follows. It is believed that these aminoacids are essential for the synthesis or efficient functioning of the enzymes thymidine synthetase and thymidine kinase which are involved in DNA replication. These enzymes are probably absent or at low concentration in the cell during G_1. The 20 per cent of DNA synthesized is at the expense of pools of compounds which are already further along the synthetic pathways than the points at which the enzymes play their part.

The fact that DNA synthesis in *Euplotes* starts simultaneously at the two ends of the macronucleus suggests that the cytoplasm plays a crucial role in the initiation of S. This is further supported by the observation that in 'giant' *Euplotes* with 2 macronuclei DNA synthesis commences at the same time at all 4 ends. It has also been suggested that nucleo-cytoplasmic interactions may be responsible for the different timing of synthesis in the macro- and micronuclei of *Tetrahymena* (Prescott and Stone, 1967).

Cytoplasmic factors are obviously very important in modifying the role of the nucleus in all cells (e.g. see Gurdon, 1970). Within the Protozoa, amoebae have lent themselves to an experimental examination of this problem. It is relatively easy to remove or transplant nuclei in these animals and the results of such work have been of wide general interest (e.g. Danielli, 1959; Goldstein, 1963). While ciliates are less ideal material for this type of work some nuclear microsurgery and transplantation has been done. For instance De Terra (1970) has shown that if an interphase nucleus of *Stentor coeruleus* (which is moniliform) is transplanted into an animal in the early stages of regeneration (in which the nucleus is normally compact) the transplanted nucleus soon coalesces and becomes compacted. The cytoplasmic stimulus for compaction needs to be present for at least one hour to start the process and must be present throughout to bring it to completion. The elongation of the compact nucleus to a moliniform one is also under cytoplasmic control. The cytoplasmic stimulus is only present at certain times during the cell cycle and a compact nucleus transferred to an interphase cell will not elongate. One in the process of elongation will be arrested by such a transplantation. De Terra also showed that DNA synthesis is under cytoplasmic control. Nuclei in G_1 will incorporate tritiated thymidine when transferred to a cell in S, and incorporation ceases if an S nucleus is transplanted into a cell in G_1. Her results further showed that the stimulus for DNA synthesis was the appearance of a substance (as yet unidentified) rather than the removal of an inhibitor.

RNA AND PROTEIN SYNTHESIS

Prescott (1960) showed that in *Tetrahymena pyriformis* incorporation of labelled adenine into RNA proceeds slowly in the first half of the interphase period but that during the second half it increases (85 per cent of total in the latter). It would seem that more, if not all of this synthesis takes place in the macronucleus, although the evidence does not entirely eliminate the possibility of cytoplasmic RNA synthesis. Recently Murti and Prescott (1970) have shown the presence of RNA in the micronucleus of *Tetrahymena*. This may indicate its manufacture there but it could have passed in from the cytoplasm.

Ammerman (1970) has obtained similar results from *Stylonichia*. If the micronucleus is indeed synthesizing RNA then it cannot be considered only as a genetic 'memory bank', important only at sexual or pseudosexual reproduction. There is some evidence of a somatic function for micronuclei in *Tetrahymena*. Wells (1961) showed that if the micronucleus was removed there were some detrimental effects including a longer generation time. The micronucleus is essential in some species for regeneration and asexual reproduction.

   Protein synthesis, as reflected by the incorporation of labelled methionine, continues at a constant rate throughout the cell cycle, at least in *Tetrahymena* (Prescott, 1960). Histone, a protein associated with DNA, may not be made throughout the cell cycle. In *Euplotes* its incorporation into the nucleus is closely linked with the production of DNA by the reorganization bands.

CULTURAL CYCLE

Many Protozoa are grown in the laboratory under more or less controlled conditions. It is possible to grow many ciliates in sterile media containing chemically defined ingredients (see p. 103). Other culture methods rely on having a simple ecosystem biased in favour of the organism required. Such cultures containing predators (e.g. *Didinium* and *Podophrya*) are easily maintained by supplying suitable numbers of prey organisms. With bacteriophageous ciliates the food may be grown separately and supplied to an otherwise sterile culture (*Paramecium* can be grown like this). However, more commonly the culture is provided with some sort of organic material (wheat grains, infusions of leaves, meat extracts, etc.) which encourages the growth of 'wild' bacteria upon which the ciliates feed. Such cultures will usually contain many fungi, flagellates and other contaminating organisms as well as the required animal and its food.

   Whether the culture is pure or crude it often passes through a number of phases. These phases can be seen most clearly in sterile monocultures, e.g. *Tetrahymena* in a solution of proteose peptone and yeast extract. After inoculation (Fig. 33) there is often a period in which no reproduction occurs. This lag phase may result from the senescent state of the animals in the inoculum (or, in special cases, is due to the times taken to excyst). Once this lag phase is over the animals grow and divide. Under constant conditions with excess food the increase in the numbers of animals in the culture will be exponential, that is the population will double each generation time. This period of unrestricted reproduction is known as the log(arithmic) phase. Eventually the rate of division will decrease. A variety of factors may be responsible for this deceleration in growth rate. The

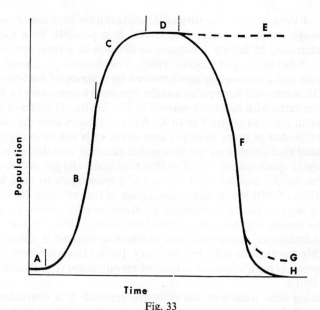

Fig. 33

An idealized representation of the growth of ciliates in culture. After inoculation there is a period of no reproduction (A) There follows a period of logarithmic growth (B) which gradually slows (C) A stationary phase (D) may result in encystment (E) or be followed by cell deaths (F) resulting in a small residual population (G) or extinction (H).

most influential will be the depletion of an essential nutrient of food organism. The accumulation of waste products may also retard growth. In crude cultures there may be competition from other organisms. After some time the numbers of animals in the culture will become constant, either because deaths are equalling growth or more likely because reproduction ceases. After this stationary phase, which may last for some time, one of a number of things may happen. Some or all of the animals may encyst; more may die leaving a remnant of survivors, or all may die. In the laboratory one usually subcultures into fresh media before this last decline is reached.

SYNCHRONOUS DIVISION

In a thriving culture there will be animals at all stages in the cell cycle, including a certain number in division. It is therefore difficult to study the biochemical changes that take place during the cell cycle. The solutions to this problem are either to confine one's examinations to single animals or to small groups of selected individuals (Prescott, 1960), or persuade all the animals in a culture to go through their cell cycle in step. The former techniques have been used to some advantage but it has severe limitations. It is possible, by a variety of techniques, to induce synchronous division in several species of ciliate (Scherbaum and Loefer, 1964). *Tetrahymena pyriformis* can be easily and conveniently synchronized by a series of heat shocks. For GL strain half-hour periods at the optimum temperature of 29 °C are alternated with half-hour periods at 34 °C. After $3\frac{1}{2}$ hours of such treatment the cells are held at 29 °C. About $1\frac{1}{2}$ hours after the end of heat treatment as many as 80 per cent of the cells will be in division. Provided that conditions are favourable the next two divisions will be largely synchronous but after that they gradually get out of step. This technique enables biochemists and physiologists to have large quantities of cells in a predetermined stage of the cell cycle. However, it may well be that cells induced to divide synchronously by such violent means will not have normal physiology. It is obvious that the technique of synchronization is effective because it affects and possibly disrupts the cell's biochemistry. Indeed this interference and disruption is itself often the subject of investigation (e.g. Byfield and Lee, 1970).

During heat treatment the cells are arrested in a characteristic phase of development (see review by Williams, 1964) with their micronuclei in an anaphase-like figure and the oral structures in the 'anarchic field' stage (i.e. the oral primordium composed of many kinetosomes but showing no organization into AZM and UM). Cells that have progressed further than this phase before the heat treatment either resorb their oral primordia and replace them with anarchic fields, or, if they are far advanced in development, they are unaffected and proceed through division normally. It is supposed that the 30-minute periods at high temperature destroy some substance necessary for development past the blocking stage. This substance has been tentatively called a 'division protein', but it is likely that more than one substance is involved. More recently it has proved possible to synchronize *Tetrahymena* by half-hour heat shocks separated, not by half-an-hour at optimum temperatures, but by the animals' generation time. About 6 or 7 such shocks are required and so this process is lengthy but it may well cause much less damage and disruption to the cells.

LIFE-CYCLES. ADOLESCENCE, MATURITY AND SENESCENCE

The life-cycle of ciliates such as *Paramecium* or *Tetrahymena* would seem to be a continuous series of binary fissions occasionally interrupted by conjugation or autogamy. This view led to the idea that such Protozoa were potentially immortal because they are pure germ plasm. Unless some catastrophe intervenes, it was suggested, such animals could continue to divide for ever. The immortality of ciliates was contested by Huxley (1926), among others, in his essay 'The Meaning of Death'. He picks on the moment at which the macronucleus degenerates prior to conjugation as 'the death of the directive portion of the soma', a death only different in degree from those animals in which the soma is relatively much larger than the germ. Life must be 'recreated in a new one (macronucleus) that the micronucleus is to make'. In a sense these ideas have been elaborated and confirmed in the work of Sonneborn, Siegel and others who have shown that as the generations go by so the state of the ciliate changes. Such changes used to be thought to reflect inadequacies in the environment, but this is not the case. Seigel (1967) has shown that many aspects of these ageing phenomena are under genetic control. The typical life-cycle* of, for example, *Paramecium aurelia,* is as follows. After conjugation the exconjugant divides a number of times. Until they have undergone the first 35 fissions the offspring are incapable of conjugation. After further divisions progeny can be induced to conjugate but cannot undergo autogamy. This is followed by a period of transition when both conjugation and autogamy can occur; in a final period autogamy only is observed. Thus the life-cycle passes through successive stages of sexual immaturity, maturity and senescence. Sonneborn and his co-workers have further shown that during senescence viable offspring are produced only by autogamies that occur early in this period. Exautogamonts from late senescence are either non-viable or their progeny die after a few generations. If senescent animals are not allowed to undergo autogamy they become abnormal in appearance, divide less frequently and their death rate rises sharply. Similar results, differing only in detail, have been obtained from other ciliates.

There is convincing evidence that this sequence of events is not the result of environmental conditions but that it is controlled endogenously. For instance the sequence appears in spite of constant cultural conditions and in the presence of adequate food supply (it is often erroneously supposed that autogamy will only occur in starving animals). The macronucleus plays an important part in the ageing process. For example if a senescent individual undergoes

*Workers in this field of ciliate genetics use the term life-cycle to indicate the period of progressive and irreversible series of phenotypic changes between one nuclear reorganization (conjugation or autogamy) and the next. This will cover many binary fissions. Elsewhere in the literature a life-cycle refers to the period from the production of one adult animal to the time when its offspring become adult. In this latter usage the span of a life-cycle of *Paramecium* would be the time from one binary fission to the next, sometimes referred to as a cell cycle.

autogamy the macronucleus is reformed and the offspring are immature. It is possible, however, to induce fragments of the old micronucleus to survive autogamy, when they will regenerate. In this case
the offspring are as senescent as the parents from which the fragments
came.

A number of experiments have shown that the ageing process is
associated with the expression or suppression of various genes. One
of the best examples comes from the work of Seigel (1963) and Seigel
and Cohen (1963) who used *Paramecium bursaria* as their experimental animal. Each ciliate produces two mating-type substances
characteristic of its particular mating type. The production of each
substance is controlled by a separate locus. It was shown that in
immaturity neither of the substances is produced. Later only one is
produced and only in maturity are both produced. Thus the process
of maturation results in the expression first of one, then of both the
loci. In *Euplotes crassus* ageing results in the expression of recessive
genes.

As animals become senescent selfing conjugation (i.e. between
animals of the same mating type) may occur; viable offspring become
rare and deaths increase. In extreme old age division rate falls and
morphological changes occur. The first phenotypic evidence of old
age in *Paramecium aurelia* is the appearance of cells with atypical
numbers of micronuclei. This may be one of the results of the large
number of mutations that occur in old age. These mutations first
appear after 80 fissions and result in an inability to produce viable
clones after 220 fissions. Cytoplasmic factors are also important in
ageing. For instance if a young cell is conjugated with an old one the
offspring of the former are more viable than those of the latter.

An interesting observation that has come from this work on ageing
is that the macronucleus has a powerful influence over the micronucleus. For example a defective 'ghost' micronucleus (little more
than a nuclear membrane) will be maintained through many generations, suggesting that the macronucleus is responsible for the production of micronuclear membrane. In cells which lack a macronucleus,
the micronucleus seems to be incapable of meiosis. On the basis of
such experiments Siegel (1967) suggests that 'age induced micronuclear defects are brought about by the macronucleus . . . via the
intervening cytoplasm'. Ageing appears to result from 'progressive
changes in macronuclear gene functions' and 'senility changes are
under genetic control'.

POLYMORPHISM

In its more usual sense 'life-cycle' is taken to mean the series of events taking place between one adult and the production of the next generation of adults. In some ciliates such life-cycles may becomplicated by the presence of one of more larval stages. In some ciliates these larvae are capable of asexual reproduction.

Sessile ciliates

Vorticella lives attached to a solid substrate by a stalk. When it divides one daughter remains on the stalk while the other is set free. This motile form, the telotroch, has a posterior trochal band of locomotory cilia. The larva swims about until it finds a suitable place to settle. At settling it loses its trochal cilia and grows a stalk. Thus *Vorticella,* in common with many sessile animals both protozoan and metazoan, has a motile, distributive larval stage. A similar situation is found in the Suctoria. In this case the stalked adult is completely without cilia. The larva is ciliated and mouthless. The method of larva production varies very much from species to species. In the simplest situation the adult simply divides transversely and the proter becomes ciliated, detaches and swims away (e.g. *Paracineta*). In *Ephelota gemmipora* external budding is more elaborate. Four to twelve protuberances develop on the surface of the animal. These receive nuclear material, become ciliated, separate and swim away. Internal budding is found in *Tokophrya infusionum*. A small slit appears in the animal's apical end which gradually extends inwards, following the contours of the surface until it carves out an ovoid piece of the adult. This piece receives its macro- and micronuciei and becomes ciliated. It moves about in the brood pouch until a pore develops through which it can escape. In *Podophrya collini* the ciliated bud is formed in much the same way, but before it separates from the adult the brood pouch everts, pushing out the larva, still attached by a narrow neck of cytoplasm. This connection soon breaks and off swims the larva. In some cases abnormalities occur in larva production. Millecchia and Ruczinska (1968) reported on a strain of *Tokophrya* in which the larvae sometimes metamorphosed into adults while still within the brood pouch. In overfed *Podophrya* metamorphosis of the larva may occur after eversion but before separation, resulting in two adults attached by a bridge of cytoplasm (Jones unpublished). Rudzinska (1961) has pointed out that an adult *Tokophyrya* has a finite life span. It passes through various stages from larva to immature adult (unable to produce larvae), mature adult and on to old age and senility. This process can be hastened by heavy feeding (gluttons beware!).

In a few species of Suctoria the whole animal can become ciliated and detach itself from the stalk. This 'escape' reaction is also found in many species of stalked peritrich.

Non-sessile ciliates

While most ciliates simply divide to produce two small adults, more complicated life cycles are not uncommon. One of the best documented and discussed is that of the apostome *Gymnodinioides inkystans* (Fig. 34). This creature was extensively investigated by Chatton and Lwoff and a stimulating review of their work and its implications in ciliate and cell biology has been produced in book form by Lwoff (1950). *Gymnodiniodes* encysts on the gills of hermit crabs. This encysted form is called a phoront. Excystation occurs

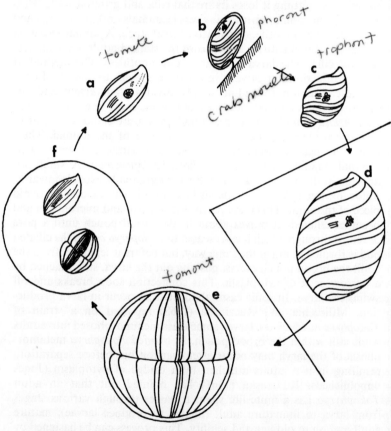

Fig. 34

The life-cycle of *Gymnodinioides inkystans*. a, tomite; b, phoront encysted on host; c, young trophont; d, growing trophont in exuvial fluid of crab; e, tomont in its cyst; f, production of tomites. After Lwoff (1950).

when the crab moults and the emerging form is called a trophont. It
feeds on the discarded exoskeleton of the crab. After 6–10 hours it
has increased its volume by up to 32 times. The trophont now es-
capes from the exoskeleton and after a few hours encysts. This stage
is known as a tomont and it divides within its cyst to produce a large
number of tomites. These break out of the cyst and live free in the
sea for 6 to 8 days. If they encounter a hermit crab they encyst on
its gills to form a phoront and the life-cycle is complete. The various
stages in the cycle have characteristic morphology and the changes
from one to another involve considerable reorganization of the
ciliature. There are also two distinct types of cysts in the life-cycle.
The phoretic cyst simply protects the ciliate while it awaits the host's
moult. The tomontic cyst encloses the animal while its undergoes considerable simplification of the ciliature and divisions to produce tomites.

Cysts of similar functions are found in other ciliates. Division
cysts are usually thin-walled while the resting cysts are thick-walled.
Members of the family Colpididae form both types. During their
time in resting cysts they usually undergo a considerable amount of
dedifferentiation including the loss of somatic and oral ciliature.
Such cysts are able to withstand desiccation. Van Wagtendonk (1955)
obtained normal excystation from *Colpoda* cysts that had been kept
in a vacuum of 10^{-6} mm Hg for a period of over 7 years. Dried cysts
of *Colpoda* will survive both high and low temperatures (100 °C for
3 hours and liquid air for $13\frac{1}{2}$ hours). Not all ciliate resting cysts are
so robust. Those of *Didinium nasutum* and *Euplotes taylori* are both
killed by drying at room temperature. Van Wagtendonk (1955) has
reviewed the factors that induce en- and excystation. The former
include a deficiency of food (*Colpoda*), an excess of food (*Blepharisma*), presence of excretion products (*Dileptus*), pH changes (*Colpoda*),
desiccation of the medium (*Euplotes*), lack of oxygen (*Tillina*) and
crowding (*Dileptus*). Excystment can be induced by osmotic phenomena (*Colpoda*), high food and oxygen concentration (*Tillina*)
and the presence of certain organic materials (*Colpoda*). However,
many of the experiments were carried out under poorly controlled
conditions and need careful repetition. A recent detailed study of
encystation in *Colpoda steinii* using modern techniques has yielded
interesting results (Tibbs, 1966; Tibbs and Marshall, 1970). They
showed that just before excystation water was expelled from the cytoplasm. This has the dual effect of increasing the concentration of
cytoplasmic solutes and decreasing the volume. The cyst wall that is
laid down is made of protein, rich in glutamic acid, which is produced in large quantities prior to encystment. There is also an
increase in carbohydrate synthesis before the wall is laid down presumably concerned with the creation of storage products.

COLONIAL GROWTH

Many peritrichs form arborescent colonies. Although there is a certain degree of specialization of zooids within the colonies they all retain the same basic morphology and feeding methods. In *Zoothamnium* colony growth has been studied in some detail (Summers, 1938a,b; and see Tartar, 1967, for review). The colony starts with the settlement of a telotroch (in this genus it is often called a migratory macrozooid). This grows a stalk and then divides to form a large and a small zooid. The larger, known as an apical zooid, continues to lengthen its stalk and to divide to produce further small zooids. These form the side branches of the colony by growth and division. The zooid at the base of each branch (the axial macrozooid) grows to a large size and is released as a migratory macrozooid capable of forming new colonies. If the terminal zooid of any branch is removed its proximal neighbour takes over its role as a divider.

8

REPRODUCTION AND GROWTH: III

MORPHOGENESIS

Morphogenesis has been defined as the 'coordinated elaboration of visible parts by which the construction or reconstruction of individual organisms of specific form is accomplished' (Tartar, 1967). In Protozoa this process occurs in a single cell and involves organelles. Studies of ciliate morphogenesis have been mostly concerned with cortical structures as these comprise the major morphological features of the animal (and also because they lend themselves to experimentation). However, morphogenesis of the new macronucleus following conjugation is not without interest.

The nucleus

During conjugation or autogamy the macronucleus is broken down and is replaced by a nucleus which derives from the new synkaryon (see p. 132 and Fig. 32). The nucleus that gives rise to the new macronucleus is called a nuclear anlage and its development has been reviewed by Raikov (1969) who divides the process into four stages. At first the anlage has the appearance of a mitotic prophase with very thin tangled chromosomes. This is called the spironeme stage. In some cases the chromosomes appear to be longitudinally split. These long threads fragment into smaller pieces which probably indicates that the long threads are not single chromosomes but a number joined end to end. These short chromosomes now enter the second stage, that of polyploidization. The chromosomes of *Bursaria,* for example, have been observed to undergo two endomitotic cycles during which they split longitudinally first to form dyads and then tetrads. After this the chromosomes either become distorted or disappear. In the suctorian *Ephelota* three endomitotic divisions have been observed. In *Nycototherus* all the chromosomes are joined into a single spironeme and the endomitotic activity results in a structure that looks like a polytene chromosome of, for instance, the salivary gland of a dipteran. In the third or achromatic stage the anlage becomes homogeneous in appearance, staining rather evenly by the Feulgen method. In most cases the only structure that stains strongly is the karyosome. There is debate as to whether the disappearance of the chromosomes at this stage represents their destruction or their despiralization. The final stage is characterized by intense DNA synthesis throughout the anlage which becomes strongly Feulgen-positive and can incorporate tritiated thymidine. The karyosome is usually resorbed at this stage. Although this DNA synthesis suggests endomitosis, confirmation is difficult due to the poor visibility and small size of the chromosomes. However, in the hypotrich *Styloni-chia* this stage is characterized by the passage along the elongated anlage of five pairs of reorganization bands (see pp. 59 and 138), each of which presumably represents a duplication of the chromosomal material. Ammerman (1965) was able to correlate the passage of the bands with a 32-fold increase in DNA concentration. Nucleoli make their appearance at this stage but up to this time the anlage contains no RNA. At the end of this stage the nucleus can be considered to be a mature macronucleus.

Stomatogenesis

When ciliates divide at least one new set of mouth parts must be produced. In some cases one of the daughters, usually the proter, inherits the parental mouthparts but often these are partially or completely resorbed prior to division. Usually stomatogenesis is finished by time of division so that both daughters have functional mouthparts immediately after separation.

The ways in which the mouthparts are produced are extremely varied. They have been classified by Tartar (1967) as follows:

(1) In ciliates with a terminal mouth (e.g. *Prorodon*) the kineties which are transected at binary fission simply turn in at the anterior end of the opisthe to form the new mouth.

(2) The mouth may be formed by the extension of two lateral kineties (e.g. the trichostome *Trimyema*).

(3) Parts of many kineties may become disorganized and the kinetosomes thus released form an oral primordium (e.g. members of the Colpididae).

(4) Parts of the few kineties may be involved. In *Lionotus* for instance the oral primordium is formed by a proliferation of kinetosomes between only two kineties.

(5) A single (stomatogenic) kinety only may supply the kinetosomes for mouthpart formation as in *Tetrahymena* (Plate 4).

(6) No kineties may be involved, the primordium being organized from an aggregation of previously scattered kinetosomes (e.g. hypotrichs).

(7) The primordium may cut across many kineties as in *Stentor*.

(8) The mouthparts may be formed by two primordia, each forming different parts of the oral apparatus (e.g. *Dileptus*).

(9) Stomatogenesis may be initiated by fibrous elements and only in the later stages involve kinetosomes (e.g. *Ichthyophthirus*).

(10) The new mouth may derive directly from the parental one, as in peritrichs.

The most complete experimental investigations of stomatogenesis are those of Tartar (1961, 1967), Weisz (1954) and others using *Stentor*. *Stentor* is a large ciliate with a marked cortical pattern and an ability to survive even extensive microsurgery. Both of these properties have been put to good use in a series of elegant transplant and regeneration experiments (which incidentally require enormous manual skill). The most commonly used surgical manipulation in these experiments is an excision of all or part of the AZM and gullet. This induces the formation of an oral primordium and eventual regeneration of the oral structures. The primordium is formed in the same place as in animals about to divide but of course in this case there is no fission. The regenerators can be further operated on, or their primordia removed and grafted into other Stentors. The species used in almost all these studies is the blue *S. coeruleus*. In this animal the oral primordium arises in the mid-ventral region (Fig. 30) and in its development splits a number of the kineties and the pigmented stripes that lie between them. This early development is initiated in the area of stripe contrast (p. 31 and Fig. 35B), a region that has special morphogenetic properties.

Only primordia in the early stages of development can be resorbed. It also seems that quite early in development parts of the primordia are determined (in the embryological sense), i.e. destined to become a particular part of the oral apparatus. Thus if the posterior end of the animal is removed together with the hinder part of the primordium (called an anlage when it has started development), then the gullet region is lacking in the regenerated animal (re-regeneration soon follows, however). If the anterior of the anlage is removed, then the gullet developes normally but the AZM is of much reduced length. Removal of all but the most posterior part of the anlage leads to a perfect gullet being formed but no AZM at all. Similar operations on very early primordia do not have these results. These primordia

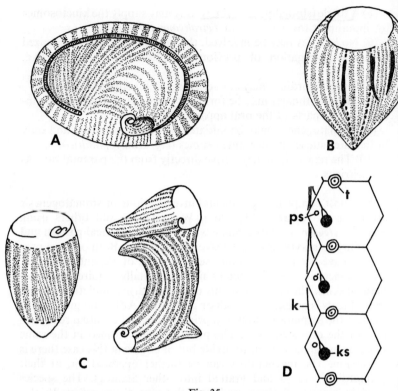

Fig. 35

A, oral field of a *Stentor coeruleus* that has undergone two reorganizations. **B,** 3 oral primordia induced in a *S. coeruleus* by the grafting of a piece of wide-stripe cortex into a narrow-stripe area. **C,** the result of reversing the polarity of the central portion of a *S coeruleus* (left) is that each part retains its own polarity and the central portion regenerates its own oral structures (right). **D,** diagram of the cortical organization of *Paramecium* showing 3 unit territories. k, kinetodesma; ks, kinetosome; ps, parasomal sac; t, trichocyst. A, B and C, after Tartar (1961); D, after Sonneborn (1970).

are not determined and complete mouthparts develop from the remaining section.

Resorption of the primordia can be induced by changing the 'cell state'. If, for instance, a small *Stentor*, regenerating its mouthparts, is grafted to a large morphostatic cell the primordium is resorbed. If on the other hand a large regenerator is grafted to a small morphostatic animal primordium formation is induced in the latter. This suggests that general cell state is of importance in the maintenance of primordia. Primordia will not be produced if there is either an inhibition or lack of activation. The change of state necessary at the beginning of regeneration probably accounts for the 3–4 hour lag between the operation and the appearance of the primordium. In activated cells this lag is only 1 hour. The lag seems to be proportional to the amount of AZM and gullet removed in the operation. Thus primordia appear a long time after minor damage but there is a much more prompt response to more drastic removals. During division this inhibition must be removed, because primordia develop in the presence of complete adult mouthparts. Once formed, a primordium inhibits the formation of a second such structure. This can be shown experimentally in *Stentor* by grafting a complete primordium on to an early regenerator. The early primordium is resorbed.

Oral primordia are formed not only before division and during regeneration but also during a process called reorganization. In this process a rather small primordium is formed which produces a membranellar band which fuses on to the end of the existing AZM. The original gullet dedifferentiates and a new one is formed at the end of the now lengthened AZM. Animals which have reorganized can be identified by the complex pattern on their frontal fields (Fig. 35A). Reorganization can be induced by first starving stentors and then feeding them (De Terra, 1970). During starvation the animal shrinks in size and the AZM is reduced. When feeding starts again the animals grow but the AZM can only increase in size by reorganization.

De Terra has attempted to explain the controlling influence behind the production of an oral primordium at division, regeneration or reorganization. She has postulated that there is a critical ratio of size of the oral apparatus to the body size (oral/somatic ratio). If a cell grows it reaches a size where this ratio is reduced below a certain value and then an oral primordium will form. Similarly in reorganizers the oral/somatic ratio becomes small, but these cells are not big enough to divide and primordium production leads not to two cells of oral structures but to an enlargement of the existing set. In regeneration the oral/somatic ratio is altered directly by the removal of all or part of the oral apparatus together with only a relatively small amount of cytoplasm. This again results in primordium production. De Terra has suggested that the important somatic factor is the size of the infraciliature. This is supported by the following experiment. Oral primordia were removed from dividing animals. The animal failed to divide but the cleavage furrow advanced far enough to transect the kineties. These animals were now double individuals as far as the infraciliature was concerned but the more posterior 'individuals' had no oral apparatus. Their oral/somatic ratios were zero and as predicted by De Terra primordia were induced in them. This effect would not have been produced if the ratio were dependent upon cell volume or total surface area.

Site of primordium formation

In *Stentor* the oral primordium usually develops at the locus of stripe contrast (see p. 31). Tartar (1961) has shown that primordium development can occur in other places. He found that a primordium site taken from one animal and grafted into the 'back' of another will develop normally (provided, of course, that the recipient's oral apparatus has been removed). In this case a primordium also develops in the normal position and the regenerated animal has two sets of mouthparts.

If a piece of *Stentor* cortex with narrow stripes is transplanted into a region of wide stripes, then a primordium will be produced at the edge of this graft in the artificially created locus of stripe contrast. Similar results are obtained if a wide stripe area is transplanted into a narrow stripe region. If there is sufficient contrast on both sides of the graft then two primordia may be initiated as well as the one in the natural position. The transplanted piece need not be in its normal orientation. Provided that there is stripe contrast there will be primordium initiation whether the graft is its normal, reverse or perpendicular orientation (Fig. 35B). Although the stripe contrast seems important in the initiation of primordia, they are formed in the narrow stripe region.

Tartar showed that the amount of stripe contrast needed for primordium formation was very small. For instance he grafted together two 'backs' which lack either very wide or very narrow stripes and found that despite this small difference in stripe width the regenerating 'animal' formed two primordia. However, longitudinal stripes of cortex containing only a few stripes would heal to produce fragments with almost no contrast. Such fragments although viable (they survived for 5 days) did not develop primordia. Tartar interprets these and other experiments as indicating an induction of the primordium in the narrow stripe region by the adjacent wide stripe region and comments that it is 'remarkable that on the cell level in *Stentor* we find something very much like induction as manifested in the embryogenesis of amphibia . . . this evocation in both cases can be brought about by the juxtaposition of certain parts, and is followed by a regionalization or secondary induction, which in Stentors is represented by mouthpart formation'. These ideas contrast with those of Weisz (1954) who regards certain kineties of *Stentor* as specialized. These 'control the polarity of the organism . . . they control the development and maintenance of probably all surface organelles (including contractile vacuoles) except the general body ciliation. If the specialized kineties are excised, some of the others may after a long interval acquire the specialized functions.' It is difficult to reconcile these statements with some of the experimental results (e.g. the production of three primordia when a region of wide stripes is transplanted into a narrow stripe area, and the production of primordia when two 'backs' are grafted together.

Cortical structures and orientation

Pieces of the cortex of *Stentor* when grafted into other stentors retain to a large extent their original polarity (Tartar, 1961). Thus if pieces are grafted into the host upside down (heteropolar) there is a conflict between the two polarities which may be resolved in a variety of ways.

(1) The graft may migrate to become homopolar. This happens even if two half animals have been grafted in a heteropolar way.

(2) The graft may migrate to the anterior of the host and be absorbed.

(3) The graft may migrate to the posterior of the host and establish a new individuality with the polarity of the graft.

(4) The graft may stimulate growth into malformed animals having perhaps two sets of mouthparts with opposite symmetries.

(5) In the case of animals in which the polarity of the middle section has been reversed, all parts retain their polarities and the mid section regenerates a new head and tail (Fig. 35C).

The transplanting of cortical grafts in *Paramecium aurelia* (for review, see Sonneborn, 1970) has produced interesting results. This animal's cortex is divided into about 4000 'unit territories' (Fig. 35D) each of which is asymmetrical. These territories are arranged in rows corresponding to the kineties. When a *Paramecium* grows it must produce new territories, indeed the number must double in each generation. It is known that all the new territories are produced in about 45 minutes just before division (which is only about 15 per cent of the generation time under optional growth conditions). All the new corticular organelles (except trichocysts) arise within unit territories, which then elongate and divide transversely. The trichocysts develop deep in the cytoplasm and move to their positions in the cortex after the new territories have been created. New territories remain in their original kinety rows and do not move sideways. Beisson and Sonneborn (1965), by pulling apart conjugants, obtained animals having small pieces of their partner's pellicle. These grafts were sometimes in a heteropolar position. The fate of these grafts was then followed through many generations. After a few divisions a few cells contained complete kinety rows of territories, showing that they were still heteropolar. This orientation was maintained even after 800 generations and survived sexual reproduction. Sonneborn's important deduction from these results is that 'the location and orientation of new parts formed during unit proliferation are determined solely *within* the existing unit and not by the polarity or asymmetry of the cell as a whole or by the orientation of immediately adjacent rows of units, or by the conventional genetic machinery of the cell'. Sonneborn (1963) was also able to create doublet animals resulting from the fusion of two conjugants. These creatures have two sets of oral apparatus and remain double through many sexual and asexual generations, showing that nuclei and fluid cytoplasm play no part in the maintenance of cortical structures. Occasional 'natural graftings' occur at conjugation when one conjugant receives some part of its partner's cortex. This graft is maintained and may result in the formation of a second set of mouthparts.

The autonomy or fixity of the *Paramecium* cortex seems to be much greater than in *Stentor*. In the latter, as has been described, cortical grafts undergo considerable modification in a single interdivisional period whereas grafts in *Paramecium* show little change even after many divisions. Regeneration in *Paramecium* can only take place at, and is indeed a modification of, the normal proliferation of territories just prior to division (Chen–Shan, 1969) while in *Stentor* the much more plastic cortex can undergo radical regenerative changes at any time in the cell cycle. It is difficult to see whether this morphogenetic rigidity in *Paramecium* is due to the pellicle being more or less independent of nuclear control or to an inability to reorganize itself during most of the cell cycle. Other ciliates are able to reorganize their pellicles to a very great extent. Hypotrichs for example undergo drastic dedifferentiation and reformation after even minor injury.

It has been suggested that the autonomy of the *Paramecium* cortex may be due to the presence of extranuclear DNA in the kinetosomes. Although there is some evidence of DNA at such sites (Smith–Sonneborn and Plaut, 1967) it is not wholly convincing. Sonneborn (1970) has pointed out that some part of the control of new structures may result from the existing structures acting as templates. However, this would not be sufficient to explain the complicated series of events that occur during the creation of a new kinetosome, nor would it explain the inductive properties of pieces of grafted oral apparatus (Sonneborn, 1963). The concept of pre-existing structures acting as templates and the idea of self-assembly of organelles from subunits are widely held in embryological circles, but they would often seem to be oversimplifications. Tucker (1971a) working with the development of microtubular organelles in *Nassula* has challeng- ed the generally held view 'that structural molecules are synthesized on ribosomes distributed throughout the cytoplasm and that these, and all other molecules influencing the growth and stability of or- ganelles, reach their sites of action by random diffusion through the cytoplasm'. He points out that something more is needed to explain the complicated morphogenetic events that occur during the repro- duction of *Nassula*. For instance it is known that microtubules can be assembled from their subunits *in vitro* provided the preparation is 'seeded 'with portions of microtubule. But Tucker argues that such self-assembly is not sufficient to explain the fact that micro- tubules can be made in one part of the cell while they are being broken down elsewhere in the same cell. In a similar way some cilia are developing, some being resorbed and yet others remain unchang- ed; some cytopharyngeal rods lengthen while others shorten. This and other evidence points to a strict localization of the various morphogenetic events in specific regions of the cytoplasm.

Development of new cilia

If has long been supposed that kinetosomes arose from other kine- tosomes by division. Thus kinetosomes could be thought of as a 'visible unit of genetic continuity' (Lwoff, 1950), a theme that has been developed and which will be discussed on p. 159. The assertion that kinetosomes divide was based on work with the light microscope. We now have the results of two relatively thorough electron micro- scope studies of the morphogenesis of basal bodies and cilia, one using *Paramecium* (Dippell, 1968) and the other *Tetrahymena* (Allen, 1969). These two reports are in essential agreement and the following description draws on both. The first indication that a new kinetosome is being formed is the appearance of a flat disc of fibrous material about $0.75\ \mu$m anterior to an adult kinetosome. This is not a part of the adult kinetosome, but for the earlier stages of its develop- ment it is loosely bound to it by dense fibrous material. Nine micro- tubules now appear in the disc. Dippell reports that they develop in an ordered sequence; first anticlockwise round the disc (when looking from the posterior of the animal) and then the last few devel- op in a clockwise sequence. This cylinder of single microtubules has its long axis at right angles to the long axis of its neighbouring adult kinetosome, and is therefore lying parallel to the animal's surface. Once 6 or 7 of the singlets are present a second round of micro- tubule proliferation occurs, converting the singlets into doublets. These second microtubules develop as C-shaped outgrowths of the singlets. Yet another round of tubule production follows to create the triplet structure characteristic of the mature kinetosome. The new basal body elongates rapidly at this time and the spoke-and-hub system appears. The distal end of the basal body has an electron-dense cap. The growing kinetosome now begins to tilt upwards towards the cell surface and the accessory cytoplasmic fibres and microtubules begin to be formed. Finally it comes to lie parallel to and about 0·5 μm from the mature kinetosome. It now has a basal granule and dense material in the lumen. Its distal end lies close to the surface and is covered by a single unit membrane. Such a basal body is now ready to sprout a ciliary shaft.

Dippell and Allen disagree as to the factors important in shaping the development of the new basal body. Allen suggests that the cart-wheel structure and the singlets may act as 'a locus of morpho-genesis'. Dippell on the other hand observes that a complete spoke-and-hub does not appear until relatively late and suggests that attributing an organizing role to it is 'putting the cart before the horse'. Rather she gives emphasis to transient 'spacers' that deter-mine the spatial arrangement of the microtubules. It is obvious that further work in this field will yield fascinating results of great im-portance to the whole field of organellogenesis. Already these studies have shown that, contrary to the claims of light microscopists, new kinetosomes arise close to, but not by division of, pre-existing kine-tosomes. There can be no doubt that the adult basal bodies are of enormous importance in the formation of new ones but as yet we have no idea of the nature of this influence. It is difficult to see how DNA (even if there is any in kinetosomes) can be passed from an adult basal body to a developing basal body. This consideration makes their 'genetic continuity' unlikely. Both Dippell and Allen found that more than one kinetosome may develop in the vicinity of an adult basal body. It is possible that the latter exerts its influence only in early stages of development. Once these are over, morpho-genesis will continue unaided and the adult kinetosome is now free to induce a second generative disc. Allen has suggested that once formed the elaborate infraciliary elements will play an important part in positioning the new cilium and establishing the correct size and position for the unit territory.

ROLE OF THE NUCLEUS IN MORPHOGENESIS

It is well established that the macronucleus is required for complete morphogenesis and regeneration (see Tartar, 1967, for review). Ex-periments with drugs have shown that if the chain of events between DNA transcription and protein synthesis is broken, then regeneration is impaired. For instance morphogenetic capacity is reduced by actinomycin-D (blocks transcription of RNA from DNA), 5-fluorouridin, aza-uracil (result in the formation of faulty RNA), RNAase (destroys RNA), p-fluorophenylalanine, puromysin and chloramphenicol (prevent transcription of protein from RNA). Removal of the nucleus acts of course to remove DNA from the very beginning of this chain. Some unicellular organisms are able to survive for quite long periods of time after enucleation, utilizing RNA already present, but ciliates are weak in this respect.

Although in many ciliates the micronucleus seems to play no part in morphogenesis, there is evidence that emicronucleation can in some species halt or reduce the rate of division (*Paramecium*, *Blepharisma*) and in others block regeneration (*Uroleptus, Euplotes*).

ROLE OF THE CYTOPLASM IN MORPHOGENESIS

We have already seen that according to various authors cytoplasmic organization probably plays an important role in morphogenesis and inheritance of cytoplasmic characteristics. It would seem that the nucleus is concerned with issuing instructions for the synthesis of the molecules needed in morphogenesis. But as Tartar asks 'Is synthesis sufficient to account for the morphogenesis?' The self-assembly hypothesis might claim that it is, and there seems no doubt that many fibre and tubule systems can 'crystallize' out from a solution of sub-units. Moreover, cilia form only in certain places and under the influence of pre-existing kinetosomes. Even in *Stentor* pre-existing structures are important. If a *Stentor* has all its cortex removed, the nucleus and endoplasm continues to live for some time but will never develop a new cortex. Surely all the mechanism for the synthesis of the necessary molecules exists; what is lacking is the structural and spatial pattern. Despite this, the answer to Tartar's question must be 'no'. In so many cases there is evidence of higher control and of co-ordinated action that seem to direct the synthesis rather than be controlled by it.

In studies of ciliate morphogenesis kinetosomes have often been credited with special properties. As we have seen they were thought to arise by division from other kinetosomes. The French proto-zoologists and cell biologists Chatton and Lwoff developed this theme fully (see review by Lwoff, 1950). They were impressed by the kinetosome's apparent genetic continuity and by its polyvalency (i.e. its ability to produce other organelles, cilia, trichocysts, etc.). Unfortunately both of these properties attributed to kinetosomes have now been shown to be largely incorrect; they do not arise by division and they do not produce trichocysts. Lwoff adopted the view of Weiss (1947) that the cell is divided into various ecological environments and that 'different species (of molecule) will automatically become segregated into their appropriate environments' resulting in a 'space pattern'. Slight changes in these space patterns, Lwoff claims, could result in kinetosome proliferation or in a tendency for the kinetosomes to produce some particular organellar material (e.g. peritrich stalk fibrils). These in turn may lead to further changes and new morphogenetic events. Lwoff regarded the kinetosome as having the capacity to organize itself and the surrounding cytoplasm, and as having a dominant place in the environment. Although recent work has invalidated many of the details of Lwoff's ideas the main themes are as stimulating as ever. Tucker (1971a) when he sees microtubules being produced in one part of *Nassula* while they are being destroyed in another may be observing the effects of such a space pattern. Allen also from his work on developing kinetosome and unit territory would probably support the suggestions that the kinetosomes play an important part in organizing the tubule and fibril systems. Others, however, have suggested that, rather than being the leader in morphogenesis, the kinetosome is dancing to a tune called by some other organizing force. Tartar (1967) has drawn our attention to inductive processes (such as oral primordium formation, see p. 154) and stated that these cannot be explained by self assembly or by kinetosomes but 'rather we can infer that the "information" resides in the surrounding cell cortex'. As yet we are ignorant as to the nature of this 'information' but as Sonneborn has written 'in addition to the beautiful world of the gene a new and wonderful world of essential and decisively directive extragenic structure is now inviting exploration'.

9

TAXONOMY AND EVOLUTION

CLASSIFICATION

In 1954 the Society of Protozoologists established a Committee on Taxonomy and Taxonomic Problems to prepare a revised scheme of classification of the Protozoa. This scheme was to take into account the results of recent research and it was hoped that it would be 'widely adopted by writers of zoological and biological textbooks' and 'serve as a valuable point of common reference for professional protozoologists'. A decade later this Committee produced its revised classification (Honigberg et al., 1964). It included a number of radical changes, including the merging of the sarcodines and the flagellates into a single subphylum, the Sarcomastigophora. For those interested in ciliates there were two major points of note. Firstly the opalinids were removed from the ciliates and placed in a separate superclass of the subphylum Sarcomastigophora. This group had previously been considered by many to be primitive ciliates and placed in a subclass Protociliata of the ciliates (Hyman, 1940, Hall, 1953). The Committee's move recognized both the opalinids' probable affinity with the flagellates and their lack of affinity with the ciliates. Secondly the subphylum Ciliophora was divided into 4 subclasses representing, as the Committee say, 'a compromise between the more conventional and the Fauré-Fremiet–Corliss scheme'. This latter had only 2 subclasses. The classification that follows is that of the Committee. The brief descriptions of each group are presented as a general guide and are not intended to be diagnostic. The genera in brackets are more or less typical of the group and have been mentioned in the text. They are not, however, 'type' genera in the strict sense. Fig. 37 shows examples of the different buccal organizations.

F

162   The Ciliates

Subphylum Ciliophora
  Class 1   Ciliatea
    Subclass (1) Holotrichia. Simple somatic ciliature. Specialized
      holophrya, puros   buccal ciliature often absent and simple when
                                 present.
    prorodon   Order I   Gymnostomatida. No oral ciliature. Cytostome
                                 opens to the outside. Cytopharyngeal wall with
                                 rods.
          Suborder (I)   Rhabdophorina. Cytopharynx with ex-
                            pansible armature of toxic trichocysts,
                            usually carnivores. (*Coleps*, Fig. 1A;
                            Dileptus, Fig. 1B; *Lacrymeria*, Fig. 1C.)
          Suborder (II) Cyrtophorina. Flattened, with ciliature and
                            cytostome on ventral surface. Commonly
                            herbivores. (*Nassula*, Fig. 2A.)
      Order II   Trichostomatida. Uniform somatic ciliature.
                     Vestibular but no buccal cilia. Often complex
                     body form. (*Colpoda*, Fig. 1D.)
      Order III   Chonotrichida. Somatic ciliature lacking in
                     adults. Vestibular cilia in apical 'funnel'.
                     Adults vase-shaped, attached to crustaceans.
                     (*Spirochona*, Fig. 2C.)
      Order IV   Apostomatida. Somatic ciliature of adults spirally
                     arranged typically with a rosette near the in-
                     conspicuous cytostome. Often parasites of
                     marine crustaceans. Have polymorphic life
                     cycles. (*Gymnodiniodes*, Fig. 3A.)
      Order V   Astomatida. Cytostome lacking. Ciliature uni-
                     form. Mostly parasitic in oligochaetes. (*Ano-
                     plophrya*, Fig. 4D.)
      Order VI   Hymenostomatida. Somatic ciliature uniform.
                     Oral ciliature composed of one UM on right of
                     a zone of 3 membranelles.
          Suborder (I)   Tetrahymenina. Oral ciliature of UM and
                            AZM. Usually small. (*Tetrahymena*, Fig.
                            2B.)
          Suborder (II) Peniculina. Oral ciliature dominated by
                            peniculi deep in buccal cavity. (*Para-
                            mecium*, Fig. 6A.)
          Suborder (III) Pleuronematina. Somatic cilia often sparse
      Scuticociliatida   but prominent caudal cilia common. Large
                                 UM. (*Cyclidium*.)
      Order VII   Thigmotrichida. Tuft of thigmotactic cilia at
                     anterior. Usually parasitic in bivalve molluscs.

Suborder (I)   Arhynchodina. Somatic ciliature uniform. Oral ciliature present. (*Hemispeira,* Fig. 2D.)

Suborder (II)   Rhynchodina. Somatic ciliature reduced. Cytostome replaced by a sucker. (*Ancistrocoma.*)

Subclass (2)   Peritrichia. Somatic ciliature absent from adults. Oral ciliature winding anticlockwise round apical pole to cytostome. Often with stalk which may be contractile.

fam: *Epistylidae* *Vorticellidae*

Order I   Peritrichida

Suborder (I)   Sessilina. Usually sessile with contractile or non-contractile stalk. Solitary or colonial. Sometimes loricate. (*Vorticella,* Fig. 4C.)

Suborder (II)   Mobilina. Motile and without stalk. Often ectoparasitic on aquatic hosts. (*Trichodina,* Fig. 4B.)

Subclass (3)   Suctoria. Adults lack cilia. Typically sessile, sometimes parasitic. Feeds through tentacles. Motile ciliated mouthless larva. (*Podophyra,* Fig. 3C.)   *Acineta*

Order I   Suctorida

Subclass (4)   Spirotrichia. Somatic ciliature often reduced. Buccal ciliature well developed. Adoral zone of many membranelles winding clockwise to cytostome. Often large.

Order I   Heterotrichida. Somatic ciliature uniform when present. Often very large.

Suborder (I)   Heterotrichina. With the characters of the order *sensu stricto.* (*Spirostomum,* Fig. 4A.)

Suborder (II)   Lichnophorina. Membranelles on oral disc. Somatic ciliature lacking. Marine ectoparasites. (*Licnophora.*)

Order II   Oligotrichida. Somatic ciliature sparse. AZM large, often extended round apical end of body. (*Halteria,* Fig. 5B.)

Order III   Tintinnida. Loricate but motile. Large membranelles. Typically marine, pelagic. (*Tintinnopsis,* Fig. 3B.)

Order IV   Entodiniomorphida. Somatic ciliature lacking. Membranelles restricted to small areas. Pellicle stiff and ornamented. (*Entodinium,* Fig. 5C.)

Order V   Odontostomatida. Somatic ciliature sparse, oral

ciliature of 8 membranelles. Small, laterally
compressed. (*Saprodinium*, Fig. 5A.)
Order VI   Hypotrichida. Somatic ciliature mostly as cirri.
AZM prominent. Dorso-ventrally flattened.
Suborder (I)   Stichotrichina. Usually many fine cirri dis-
tributed in rows. (*Stylonychia*.)
Suborder (II)   Sporadotrichina. Few, heavy cirri arranged
in 6 rows. (*Euplotes,* Fig. 5A.)

Although this scheme is generally accepted, some ciliate specia-
lists incline to the more radical Fauré–Fremiet–Corliss scheme.
Perhaps the most satisfactory presentation of this classification is to
be found in Corliss's paper of 1956 (which varies in minor details
from that put forward in his book of 1961). Corliss argues that new
techniques have so increased our knowledge of ciliate comparative
morphology and morphogenesis that a reshaping of the subphylum
is required. His proposed scheme differs greatly from earlier ones
(e.g. Hall, 1953) and unlike the Committee on Taxonomy he justifies in some detail the suggested changes.

The first major alteration (as with the Committee, of which he was
a member) is to remove the opalinids from the subphylum. Because
of their mouthless state and a number of other features they had
been regarded as primitive ancestral ciliates but Corliss (1955)
regards them as specialized and highly evolved. Their type of binary
fission, monomorphic nuclear apparatus and the possession of true
gametes are sufficient grounds to exclude them from the ciliates and
to suggest a link with the flagellates. With the removal of the opalinids the subclass Protociliatia disappears and there is no need for a
subclass named Euciliatia. Thus Corliss proposes a single class, the
Ciliata, containing all the ciliates within a single class: a move which
emphasizes the close relationships and monophyletic origins of the
group.

Corliss recognizes that the holotrichs are 'simpler and less highly
differentiated than the spirotrichs and very likely preceded them
directly in the evolutionary timetable' and consequently separates the
two groups at subclass level. This is a move which the Committee,
in proposing 4 subclasses, was not able to endorse. Corliss places
the suctorians and the peritrichs in the Holotrichia. In earlier schemes
they had been placed in a separate class mainly because of the
alleged uniqueness of the adult stage. However, Corliss's suggestion
is supported by the work of Guilcher (1951) on the mobile larva.
This work also showed that features of the adult such as stalk,
tentacles, lorica and means of reproduction have precedents in other
holotrich orders. Guilcher's work, based mainly on infraciliature

patterns, showed that suctorian larvae have affinities with the rhabdophorine gymnostomes. It is likely that these two groups evolved from a single stock. The peritrichs, like the suctoria, have specialized adult forms which resulted in their separation from the main body of the holotrichs. Corliss makes a strong plea for their inclusion in the subclass on two grounds. Firstly it is claimed that the oral ciliature is homologous with the hymenostome UM–AZM system especially when modifications of that system in the thigmotrichs is taken into account. Secondly the migration of the oral apparatus to the 'upper' (actually posterior) pole of the cell is seen as homologous to the similar movement of thigmotrich mouthparts. Corliss thus argues that the peritrichs belong to the hymenostome–thigmotrich line; arguments presumably not sufficiently persuasive to convince the Committee.

Corliss also proposes a number of minor changes from earlier schemes most of which were accepted by the Committee. In his examination of the classification, he also finds it possible to construct an evolutionary tree. The following review of Corliss's ideas reveals the roots of such a tree.

Although it was often argued that the astomes, because of their mouthless condition, are the most primitive ciliates (next to the opalinids) it now seems certain that their lack of mouths is a secondary condition. This being the case the simplest ciliates are undoubtedly to be found among the gymnostomes. It is likely that the symmetrical forms with apical mouths and no oral ciliature, such as *Holophyra,* represent the most primitive ciliates alive today. Following Fauré-Fremiet (1950), Corliss divides the group into two suborders on the basis of their cytopharyngeal structure, a more natural division than older ones based on the position of the mouth. The rhabdophorines would seem to be less advanced than the herbivorous cyrtophorines.

Guilcher (1951) not only produced evidence for suctorian-gymnostome affinities, but also showed a probable link between the cyrtophorine gymnostomes and the chorotrichs. As with the suctoria the work concentrated on the motile larva, studies of which revealed many cyrtophorine features.

The order Trichstomatida is considered to have a polyphyletic origin and is composed of 15 rather heterogeneous families. In this group, although true oral cilia are absent, the somatic cilia around the mouth are specialized. At the time of division some of the more complex members of this order dedifferentiate in a division cyst and at this time according to Corliss 'conditions typical of the simplest gymnostomes are recapitulated'. At least one of the families seems to have affinities with the cyrtophorines.

Corliss regards the order Hymenostomatida as being of central importance when considering the evolution of the ciliates. Since Furgason (1940) recognized that *Tetrahymena* had a true AZM, it has become increasingly apparent that this group exhibits the simplest form of true oral ciliature, homologies of which can be traced in more advanced groups. One of the more exciting discoveries in the study of ciliate evolution was the discovery of a 'missing link' in the form of *Pseudomicrothorax*. This creature had been placed in the Trichostomatida but examination of silver-stained preparations showed that it has the tetrahymenid arrangement of 3 membranelles as well as a structure that looks very like a UM. As well as this it has what appeared to be a cytopharyngeal basket composed of trichites allying it with the gymnostomes (Corliss, 1958a; Thomson and Corliss, 1958). The type of trichites, body shape, position of cytostome and stomatogenesis suggest affinities with the cyrtophorines, although Corliss (1961) places it in a group of unassigned tetrahymenid hymenostomes. It has further been suggested (Corliss, 1958b) that the rosette of the apostomes arose from a cytostome-cytopharyngeal complex such as that possessed by *Pseudomicrothorax*. This ciliate may thus be a link between the gymnostomes below and the hymenostomes and apostomes above. Corliss follows Fauré-Fremiet's suggestion and divides the hymenostomes into 3 suborders differing mainly in the type of oral ciliature. The Tetrahymenina contains the least specialized animals while the Peniculina have a more complex organization, especially in having a peniculus. This is a large compound ciliary organelle thought to be homologous with a single membranelle of the AZM. Members of the third suborder, the Pleuronematina, on the other hand, have an enlarged UM and a reduced AZM. Certain genera of thigmotrichs strongly resemble pleuronematina hymenostomes. Fauré-Fremiet has suggested an homology between thigmotrich mouthparts and the typical hymenostome arrangement, so it would appear that these two groups are closely related.

There seems no doubt that the Astomatida is a polyphyletic order. Indeed Puytorac (1954) has suggested that it be disbanded. He points out that many of its members are aberrant apostomes or thigmotrichs. Kozloff (1954) has even transferred an astome to the tetrahymenids on the grounds of its somatic ciliature. Corliss is of the opinion that the long evolutionary separation of these animals justifies retaining their order. It is also likely that not all the members could be reassigned and thus one would still be left with a 'dustbin' order.

Thus holotrichs can be seen to exhibit a discernible evolutionary pattern. The spirotrich orders can also be organized into a family tree.

Indeed the subclass seems to be a tight-knit group; 'the great diversity in fundamental features which is so prevalent among holotrichs appears to be absent' (Corliss, 1956). The AZM is strongly developed and the somatic ciliature often reduced.

The AZM of heterotrichs is of course homologous with that of the lower orders. In this order the somatic ciliature is often retained. In the Oligotrichida, however, the somatic ciliature is often reduced or absent. Nonetheless it is generally accepted that the oligotrichs arose from a heterotrich stock. The large order Tintinnida also seem to be related to these groups. Within the group the lorica shows a number of interesting evolutionary trends and fossil loricas have been recovered, the only fossil ciliates known (Kofoid, 1930; Campbell, 1954). Corliss also links the Entodiniomorphida with the oligotrich branch of the evolutionary tree, although the evidence is scarce because of the extreme specialization shown by the members of this symbiotic group.

With their rigid pellicle, well developed AZM and cirri to replace the somatic ciliature, the hypotrichs are usually regarded as the pinnacle of ciliate evolution. Despite the absence of clear evidence this group is also thought to have arisen from a heterotrich stock. After the suggestion of Fauré-Fremiet (1961) the order was split into two suborders. The enigmatic Odontostomatida is much in need of systematic study as little is known of its infraciliature or development. At present its affinities with other orders are a mystery.

On the basis of the available evidence, Corliss has presented us with a genealogical tree for the ciliates (Fig. 36). He falls into the trap as often happens with such trees, of drawing branches through groups which perhaps in reality sprang independently from a common stock. He also seems almost over eager to bring back into the fold long independent groups, emphasizing the similarities and ignoring the all too real differences. Nonetheless this tree is a stimulating one and makes good sense of the data. The whole field of protozoology owes a debt of gratitude to Corliss for bringing a fiery zeal to the study of systematics, setting alight a branch of zoology which is so often dull, dreary and, in consequence, neglected. The last twenty years has seen a remarkable rationalization of ciliate taxonomy and phylogeny.

THE SPECIES PROBLEM

The concept and the reality of the species are important to the biologist. Most biologists accept the two major features, constancy and sharp delimitation, that characterize a species. The former allows one researcher to compare his results with those of others working

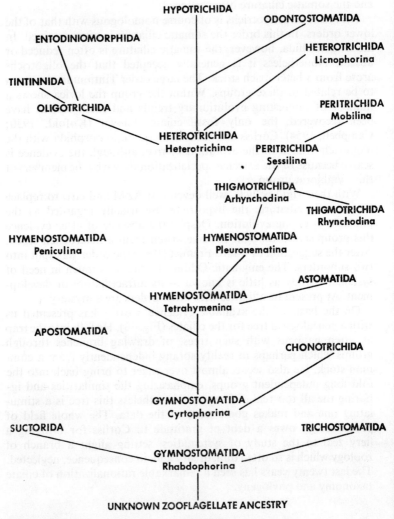

HYPOTRICHIDA

ODONTOSTOMATIDA

ENTODINIOMORPHIDA

HETEROTRICHIDA
Licnophorina

TINTINNIDA

OLIGOTRICHIDA

PERITRICHIDA
Mobilina

HETEROTRICHIDA
Heterotrichina

PERITRICHIDA
Sessilina

THIGMOTRICHIDA
Arhynchodina

THIGMOTRICHIDA
Rhynchodina

HYMENOSTOMATIDA
Peniculina

HYMENOSTOMATIDA
Pleuronematina

ASTOMATIDA

HYMENOSTOMATIDA
Tetrahymenina

APOSTOMATIDA

CHONOTRICHIDA

GYMNOSTOMATIDA
Cyrtophorina

SUCTORIDA

TRICHOSTOMATIDA

GYMNOSTOMATIDA
Rhabdophorina

UNKNOWN ZOOFLAGELLATE ANCESTRY

Fig. 36
A suggested evolutionary tree for ciliate orders and suborders. After
Corliss (1961)

with the same species (allowing for biological variation). The latter brings to biology the realization that one can deal with animals not just as individuals but as members of a group that has its own characteristics, different from those of other species. This concept of the clearly defined species, first put forward by Linnaeus, has been challenged, especially by some early supporters of Darwin who realized the necessity of a certain inconstancy to permit the evolution of one species from another. In a brilliant essay Mayr (1957) has reviewed the history of the species concept and recognizes three approaches to it. The first is based on morphological differences and, of course, is inadequate when dealing with different life-cycle stages or sexual dimorphism. The second relies on relative differences between the species. For example, one species will not interbreed with another. He likens this usage of 'species' to that of a word such as 'brother'; it does not imply a property (e.g. hardness etc.) but a relationship. The third sees the species as biological unit the members of which cannot actually or potentially interbreed with members of a different unit or species. This last approach seems to have been the most rewarding and the majority of the more recent definitions are similar to that of Mayr that species are groups of actually or potentially interbreeding natural populations which are reproductively isolated from other such groups.

As many have pointed out, including Mayr himself, such a definition is not very satisfactory when considering Protozoa. Sonneborn (1957) has written 'in obligatorily self-fertilizing organisms, this concept of the species contracts to include but a single organism. In exclusively asexual organisms the concept admits of no species at all. On this concept perhaps half or more of the species recognized by the students of the Protozoa are invalid and, in principle, cannot be modified so as to become valid.'

In some ciliates sexual behaviour is unknown or impossible (amicromuleate strains) and self-fertilization is found (autogamy) although there is no ciliate for which it is the only type of sexual activity. However the most difficult 'species problem' concerning ciliates arises over mating varieties. These are, after all, reproductively isolated units and the problem is most difficult to resolve. For example in *Paramecium aurelia* (Preer, 1969) there are at least 14 varieties. Paramecia will only reproduce sexually with members of compatible mating strains from their own variety. The varieties vary in their geographical distribution, morphology and physiology but none of these differences are, on their own, diagnostic. Should these varieties be designated species and given individual specific names? Sonneborn (1957) argues that at present it is impossible to identify the various varieties without a great deal of specialist skill, equipment and time. This is undoubtedly an argument of expediency but Sonneborn contends rightly or wrongly that 'assigning species names to varieties of *P. aurelia* is indefensible because it is totally impractical' and further claims that 'even the most devoted adherent to the biological species concept would probably agree' [with this conclusion]. He next considers the more fundamental question as to whether or not the varieties *are* species even though they are not named. He argues that the species concept based on a common gene pool is limited in many ways, especially in the cases of asexual and self-fertilizing animals but this does not invalidate the usefulness of the concept when faced with difficult cases such as that of *P. aurelia*. There would seem to be three alternatives. (1) Accept a double standard for the word species, applying it sometimes to common gene pool units and at other times using it in a routine taxonomic way (i.e. as in the case of asexually reproducing animals and where there is no evidence for a common gene pool). (2) Restrict the term species to those cases where a common gene pool has been established and find a new term for the others. (3) 'Continue to use the term species in its universal pigeon-hole sense and find a new term for the limited number of cases in which common gene pools are known or will be discovered.' Sonneborn chooses the third alternative and proposes the new term syngen for organisms capable of 'generating together'. This is a strange decision, for if one accepts the concept of a species as defined by Mayr, then surely the 'variety' (in the sense used by Sonneborn, see p. 135) in this case is homologous with the 'species', and it is the supra-varietal group that should receive new terminology. On Sonneborn's argument those groups of animals with the strongest claim to the term species, that is those that have been shown to be reproductively isolated, will be the very groups to be renamed syngens. Taken to its limit where all sexually reproducing animals are assigned to syngens, the term species would only be applied to asexually reproducing animals and the concept of the species would have been turned inside out. To propose fresh nomenclature here would merely cause irritation and confusion. At the theoretical level it would seem sensible to recognize the right of the varieties of *P. aurelia* to species status, regarding *P. aurelia* as something approaching a genus. Botanists, more used to dealing with closely related groups of species often difficult to distinguish apart might suggest calling the group of varieties '*P. aurelia* aggregate.' Many might agree that there is little practical sense in giving each variety a full binomial nomenclature though there would be little difficulty in so doing as they already have been assigned numbers.

Sonneborn's conclusions have been contested by other zoologists. Preer (1969) admits that 'mating groups actually constitute distinct biological species' but while agreeing that although it is accepted procedure to give even very similar (sibling) species binomial names, 'for practical reasons this course has not been followed in the Protozoa'. Hairston (1958) in a spirited discussion argues that the varieties should be given Linnean names. He does so mainly on philosophical grounds and claims that Sonneborn would 'deny that evolutionary affinities . . . should always be considered in reaching decisions about the catalog of nature' and that to 'deny the goal of understanding, however remote or seemingly unattainable, is to deny the existence of a science'. He claims that the difficulties of identifying the varieties have been magnified and are no greater than in some other fields of zoological taxonomy. To date his advice has not been followed to the full but most critical workers state which variety they use, giving its number; *de facto* recognition at least.

EVOLUTION OF THE MACRONUCLEUS

There seems to be no precedent or parallel in any other animal for the type of nuclear dimorphism found in the ciliates and its origin used to be regarded as a mystery. Recent work with primitive ciliates living interstitially in sand has thrown some light on this problem (see review by Raikov, 1969). These lower holotrichs fall into two types, those without nuclear dualism and those with diploid macronuclei. The former contains a single genus, *Stephanopogon,* a rhabdophorine gymnostome. *S. colpoda* grows without cell division (trophont stage) and then undergoes multiple fission in a division cyst. Small trophonts have two identical nuclei but these undergo mitoses as the trophont grows. The mitoses are usually synchronous and large trophonts contain 12–16 nuclei. In the division cyst the cytoplasm divides, cutting the animal into a number of small parts (tomites) each with a single nucleus. The tomite nucleus divides once before excystation. Raikov likens the nuclei of *Stephanopogon* to those of amoebae. Unfortunately sexual reproduction has not been observed in this genus. Raikov regards this homokaryotic condition as primary and assumes that nuclear dualism arose within the class Ciliatea.

Lower holotrichs with diploid macronuclei are much more common than homokaryotic types. About 100 species have been described including all the members of the well known freshwater genus *Loxodes*. All these animals have more than one macronucleus, the minimum is two (with one micronucleus) but usually there are more of both, up to several dozen. The micronuclei look like those of other ciliates. They are comparatively small, usually not more than 10–15 μm in diameter and their form is simple. They have relatively little DNA but are rich in RNA. Raikov has shown that in *L. magnus* a macronucleus contains twice as much DNA as a micronucleus and concludes that the former must be diploid. These macronuclei never divide but are derived from the division of micronuclei (Fig. 7). It is estimated that macronuclei degenerate after 4–7 generations. There is some debate as to whether or not these macronuclei represent a primitive state as compared with the polyploid macronuclei of other ciliates. Fauré-Fremiet (1954) regards them as secondarily simplified from the poloploid state and being a specialized adaptation to the psammophilic way of life. Raikov on the other hand regards the diploid macronucleus as primitive and 'phylogenetically more ancient than the polyploid macronuclei'. (Grell, 1962, has called them 'karyological relicts'). Raikov counters Fauré-Fremiet with the following arguments. Firstly these nuclei are found only in primitive ciliates, mostly rhabdophorine gymnostomes. Secondly diploid macronuclei are found in a large number of different families and thus presumably the condition reflects a previously more common state. It is difficult to see how convergence could have resulted in such similarity of nuclear behaviour and organization. Thirdly interstitial sand fauna contains primitive members of other phyla and this habitat seems to favour the survival of ancient forms. Fourthly, diploid macronuclei are found in the freshwater ciliate *Loxodes* and so cannot be a specialized feature associated only with a psammophilic way of life. Lastly, it is difficult to imagine how a diploid nucleus could arise by the simplification of a polyploid one.

Raikov sees the evolution of nuclear dualism in the following way. At first homokaryotic multinucleate forms arose (genera showing this condition exist in many groups of protozoa, e.g. *Actinosphaerium, Pelomyxa* and *Opalina*). The surviving example among the ciliates is *Stephanopogon*. Dualism resulted from the production of genetically identical nuclei some of which were somatic in function and others generative. The generative nuclei remained totipotent but the somatic nucleus lost its ability to divide. Such is the situation in those ciliates with diploid macronuclei. The third and final stage of nuclear dualism is that of animals with generative nuclei and a large highly polyploid macronucleus. Polyploidy in a sense replaces the multiplicity of diploid macronuclei (Polyploid macronuclei are usually single). Ability to divide is re-acquired at the same time as polyploidization. Raikov sees the non-dividing diploid nucleus as a stage in the evolution of the polyploid nucleus. Perhaps it is more likely that both arose from some other, now vanished, form of nuclear dualism in which the diploid macronucleus had not lost its ability to synthesize DNA and divide.

THE ORAL REGION

analogy:

Reference has already been made to evolutionary trends in oral ciliation (see pp. 29 and 165). Homologies have also been recognized in the non-ciliary structures and cavities associated with the mouth and these have been summarized and defined by Corliss (1959). He sees the simplest situation as that of the gymnostomes (Fig. 37A,B) where the cytostome is on the surface of the animal and there is no elaboration of the surrounding pellicle. The cytostome is defined as the true mouth 'followed only by the cytopharynx'. The first elaboration of the oral region is found in advanced cyrtophorine gymnostomes (Fig. 37C). The cytostome is now at the bottom of a depression, the vestibulum. In trichostomes the vestibulum is lateral in position and better developed (Fig. 37D). In the hymenostomes the depression around the cytostome contains the oral cilia, AZM and UM, and is therefore termed a buccal cavity. The opening into this cavity is the buccal overture (Fig. 37E). In the peniculine hymenostomes there is a vestibulum with its somatic cilia as well as a buccal cavity with its oral cilia (Fig. 37F). In the peritrichs the oral ciliature has grown beyond the confines of the buccal cavity and encircles the anterior end of the body (Fig. 37G). The buccal cavity of hypotrichs is very large and often rather flat and again the oral cilia may extend out beyond the buccal cavity (Fig. 37H).

The realization of these homologies and the consequent reduction of terms have brought order to a branch of ciliate morphology where previously similar names had been used for different structures in some cases, while in others the same structure had been given a plethora of different names. Even in the same organism there had been a bewildering variety of names for the same structure. For example the buccal cavity of *Paramecium* has been variously referred to as cytopharynx, pharynx, oesophagus and gullet.

THE KILLER TRAIT AND THE EVOLUTION OF ORGANELLES

Certain strains of *Paramecium aurelia* are known to release a toxin into the medium which will kill other strains of *P. aurelia*. The former are called killers and the latter sensitives. This property of killers attracted the attention of geneticists (see Beale, 1954; Sonneborn, 1959) as this property is inherited in an unusual way. Sonneborn established that killers of one particular stock had the dominant allele, *K,* while the sensitives had the recessive, *k*. Heterozygotes (*Kk*) were killers if they inherited their cytoplasm from a killer and sensitives if they received it from a sensitive parent. If the heterozygotes undergo autogamy some would become recessive homozygotes

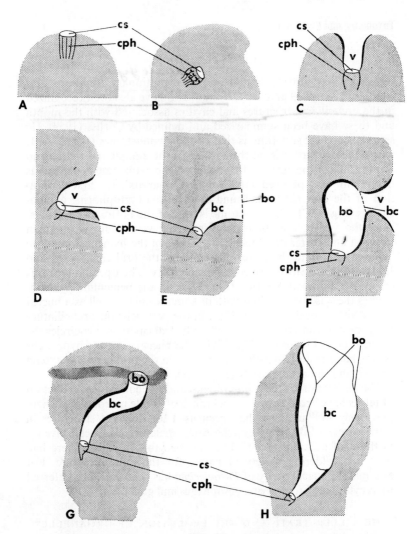

Fig. 37

Oral structures considered to be homologous. **A,** rhabdophorine gymnostome; **B,** cyrtophorine gymnostome; **C,** simple trichostome or chonotrich; **D,** advanced trichostome; **E,** tetrahymenid hymenostome; **F,** peniculine hymenostome; **G,** peritrich; **H,** hypotrich. bc, buccal cavity; bo, buccal overture; cph, cytopharynx; cs, cytostome; v, vestibule. After Corliss (1961).

(*kk*) and thus become sensitives even if their cytoplasm came from a killer. Reintroduction of the dominant *K* failed to restore killer properties to such animals. However, if a killer and a sensitive are conjugated together (a technically difficult but not impossible experiment) and cytoplasmic exchange occurs then both exconjugants are killers. The cytoplasm of killers was found to contain particles absent in sensitives. These particles are called killer particles. It was shown that for an animal to be a killer it must have two genetic determinants, a nuclear gene and a cytoplasmic factor. Killer particles are now known to be of different types and each is designated by a Greek letter, kappa, mu, etc. Later work showed that these particles appear to be more or less similar to bacteria (Preer, 1969; Preer *et al.*, 1972). They look like bacteria in both light and electron microscope. One type (lambda) can reproduce outside the paramecium, while another, kappa, has been shown to respire and utilize glucose *in vitro*.

These bacterium-like particles probably supply some nutrients to their hosts; folic acid in the case of stock 299 lambda particles (Van Wagtendonk *et al.*, 1963). There is therefore a close symbiotic relationship. The evolutionary significance of these particles is that they have practically become organelles of the paramecia in which they live. This is of particular relevance to the suggestion put forward that mitochondria and chloroplasts have arisen from prokaryotic symbionts. This is not a new idea but it is undergoing a vigorous revival at present because a good deal of supporting biochemical evidence has been obtained (see Raven, 1970; Margulis, 1971). It would seem likely that killer particles are at a stage intermediate between organelles and parasites. The fact that there are many types of killer particle suggests that such relationships are not difficult to establish.

Ciliates have recently been used to show that certain genetic factors reside in the mitochondria (presumably controlled by DNA which survives from the time when they were fully autonomous). Resistance to the drug erythromycin is one such factor. Paramecia which are resistant to this antibiotic will transfer their resistance to sensitive animals only if there is cytoplasmic mixing at conjugation (Beale *et al.*, 1972) and no effect of nuclear genes on drug resistance could be detected. If an animal with mixed cytoplasm is placed in drug-free medium its offspring may gradually lose resistance suggesting that there is some selection in the cytoplasm, with the 'wild-type' factor having an advantage (Adoutte and Beisson, 1972). Beale and his coworkers fractionated resistant cells and injected the various fractions into sensitive animals. Of 64 animals injected with a fraction largely composed of mitochondria, 54 became resistant. None of those injected with pre- or post-mitochondrial fractions became resistant. These workers were also able to show differences in the mitochondrial ribosome proteins from resistant and non-resistant animals. Such experimental examinations of cytoplasmic genetic material all add weight to the theory that mitochondria once had an independent existence.

I O

CILIATES OF ECONOMIC IMPORTANCE

SEWAGE TREATMENT

Probably the only industrial process to utilize ciliates to any degree is that of sewage processing. There are two main methods commonly employed in advanced countries; the percolating filter and the activated sludge systems. Both depend upon a variety of micro-organisms, including ciliates, for their success.

The percolating filter (see Bruce, 1969, for an introduction to this subject) was discovered in 1888 and became widely accepted by 1908. It is still widely used in the United Kingdom especially in sewage disposal works serving populations of 100 000 or less. The process is as follows. Raw sewage is fed into settling tanks in which most of the solid material sinks to the bottom. This semi-solid material is then passed to an anaerobic digestion tank in which the bacteria are the important organisms. Microbial activity in these tanks yields methane which can be used to power electrical generators whose output can meet the electrical power needs of the plant. The liquid from the top of the settling tanks, rich in dissolved and suspended organic material, passes to sprinklers that allow it to trickle on to the filter beds. These beds are a familiar sight with their sparge pipes rotating (in the case of circular filters) or travelling back and forth (in the case of rectangular filters) over the beds. These beds contain lumps of clinker, rock or slag which has a surface area of 80–110 m^2 per cubic metre of bulk volume. They are about six feet deep and the liquid drains through the filter and out at the bottom via underdrains (which also allow ventilation of the beds by natural draught). The beds may cover a considerable area; the works at Minworth (Birmingham) which serves over one million people has about 42 acres of filters.

On the surface of the clinker is a microbial association referred to as a film. This contains solid and colloidal material from the waste and a wide variety of bacteria, fungi, protozoa and higher animals (fly larvae, oligochaetes, nematodes). Algae may also grow on the upper parts of the filter which are exposed to light. As the waste liquid passes through the filter the organic material is removed. The kinetics of removal in relation to the depth of the filter traversed usually approximate to those of a first order reaction. Thus most of the material disappears in the upper layers; as much as 90 per cent of the biodegradeable matter may be removed in the upper two feet of the bed. The effluent from such beds contains solid material, usually pieces of film, which is allowed to settle before the supernatant is discharged. The exact details of processing vary from plant to plant, e.g. a common procedure is for the effluent to pass through a second bed.

In the activated sludge process the sewage, after at first settling, is passed into large open tanks in which it is vigorously agitated and aerated. In such tanks there develops a complex association of micro-organisms which live in and on small suspended lumps or flocs. As in the filter bed the micro-organisms oxidize the organic material and grow to create more flocs. These can be settled out at the end of the process and either used as fertilizer or passed back into the system to be autodigested. This method was first perfected in 1913 and is now in use throughout the world. In this country it is used especially in plants serving large communities; 7 out of 8 largest (in terms of people served) works are of this type. Smaller plants of this type are being used experimentally as a means of dealing with the large quantities of slurry resulting from modern methods of intensive cattle rearing.

In both filter and sludge processes the micro-organisms are playing a similar role. They clarify the water by flocculation and they oxidize organic materials. Protozoa would seem to be of importance in flocculation. For example Curds (1963) showed that *Paramecium caudatum* could flocculate Indian ink particles by two means. It secreted a carbohydrate, probably polysaccharide, into the medium which adsorbed on to the particles, changing their surface charge and allowing them to aggregate; secondly particles were ingested and excreted by the animal in pellets bound together with a muco-protein. The ability to flocculate particles has been observed in other species (including those peritrichs commonly found in activated sludge). It is also likely that micro-organisms other than Protozoa can cause particles to agglutinate in this way.

The second and perhaps more important function of the ciliates is that of removing bacteria and particulate organic material from the waste. Curds and Cockburn (1968) built laboratory-scale activated sludge plants. They could run these plants free of protozoa, but containing sewage bacteria. When the effluent from such a plant was tested it was found to be turbid and to contain large numbers (over 100 million/ml) of bacteria. They then added to some of the plants a large or a small inoculation of ciliates. Those with the large inoculum quickly produced an effluent of high quality containing less than 10 million bacteria/ml. The plants with the smaller inoculum took longer to reach this stage. There was no doubt that the ciliates contributed significantly to improving the quality of the effluent. Certainly ciliates can consume vast numbers of bacteria. Curds and Cockburn (1968) have shown that a single *Tetrahymena* can consume 7200 bacteria in 12 hours. As ciliates are present in healthy sludge at densities of 50 000 per ml there can be little doubt that they are responsible for keeping down the number of bacteria in the effluent.

When settled untreated sludge is first aerated it contains few protozoa and takes some time for the 'climax' fauna and flora to develop. Curds (1966) followed the way in which such sludge matures in a laboratory scale plant. The first Protozoa to appear were flagellates, such as *Peranema,* usually associated with decaying organic matter. After a few days ciliates began to appear in increasing numbers and flocs began to appear. These ciliates were mainly free-swimming forms such as *Uronema* and *Paramecium.* These are soon replaced by mainly crawling ciliates living on the surface of the flocs (e.g. *Aspidisia*). After 3–4 weeks peritrichs appear in large numbers and thereafter they seem to live in a competitive relationship with the crawling hypotrichs, resulting in alternating population peaks for the two groups. This temporal sequence in sludge is translated into a spatial one in filter beds. At the top, where the highly polluted waste enters, flagellates are common. Further down are free-swimming ciliates while the peritrichs and crawling hypotrichs are still nearer the bottom. It would appear that the 'healthiness' of a sludge or filter can be diagnosed by examining its protozoa. A large number of peritrichs indicates that the system is working well, whereas large numbers of flagellates may indicate excessive pollution with organic material. Some species of protozoa are looked upon as indicators (Curds and Cockburn, 1970). However, they point out that many species are able to tolerate a wide range of environmental conditions. One must also be careful in identifying the so-called indicators, as closely related species may occur in very different conditions (Curds, 1969). *Vorticella campanula* indicates a healthy sludge whereas *V. microstoma* occurs in polluted conditions. Despite these reservations, however, 'the presence of large numbers of active attached and crawling ciliates normally indicates a healthy sludge' (HMSO, 1968).

PARASITES

There is only one pathogenic ciliate parasite of man, the trichostome *Balantidium coli*. The genus has many species found living as parasites or commensals in a wide variety of vertebrate and invertebrate hosts. One species is found commonly in the rectum of the frog and is a convenient source of material for examination. *B. coli* inhabits the large intestine of pigs where it appears to do no damage and one must assume that the pig is its natural host. However, it can also infect a variety of primates, including man, in whom it causes balantidial dysentry. In pigs the ciliate lives in the lumen of the large intestine but in man it invades the mucosa, producing irritation which results in diarrhoea. In advanced cases blood vessels of the gut are damaged and bloody dysentery results. Diagnosis is simple as either free or encysted *Balantidium* appear in the faeces. Treatment with drugs is usually effective.

The life-cycle of *Balantidium* is simple. In the gut the ciliate reproduces by binary fission and conjugation. It is able to encyst and the cysts, passed out in the faeces, are resistant. If they are eaten they excyst in the gut. If the host is suitable they will reproduce. Man is usually infected from pig faeces. Prevention of infection is a matter of elementary hygiene and, as Baker (1969) delicately puts it, 'the avoidance of a diet which includes pig faeces'.

There are three important ciliate parasites of fish (Bauer, 1961). They are the cyrtophorine gymnostome *Chilodonella,* the tetrahymenid *Icthyophthirius* and the mobilinid peritrich, *Trichodina.* They cause a good deal of mortality in culture ponds where the high host density favours epizootics. In natural waters infection does not usually lead to more than slight weight loss.

Chilodonella infects the gills, fins and body surface and causes heavy mortality, especially of young fish in the winter months. At this time the epithelium of the fish undergoes considerable destruction favouring the development of the parasite. It has been reported that 90 per cent of winter losses in carp ponds may be due to chilodonasis. The weak fish, if not removed, are the first to be infected and allow the parasite population to rise to levels at which even healthy fish succumb. In Russian carp ponds *Chilodonella* has a temperature optimum of 5–10 °C although division does take place at temperatures as low as 1 °C. At 20 °C the parasite dies in large numbers. It thrives in shallow ponds and deep winter ponds are recommended as a control measure. Other means of control include the drying out and disinfection of the winter ponds during the summer; using a spring water-supply (to reduce the chances of entry of the parasite from outside sources), and the treatment of fish with dilute salt, formalin or acetic acid solutions.

Icthyophthirius is a rather more specialized parasite. Small free-swimming stages invade the skin of fishes and burrow about underneath, sometimes collecting in common chambers. Their movement gradually ceases and they then grow to their full size (up to 1 mm long). When fully grown they break out of the host and encyst attached to aquatic vegetation or other submerged material. Within this cyst many binary fissions take place yielding up to 2000 swarmers each about 30–40 μm long. The cyst opens and the swarmers swim in the water until they come in to contact with a suitable fish host. Neither the swarmer nor the adult can survive more than a few days outside the host (except in the division cyst). The parasite (at least the species found in carp ponds) has a high temperature optimum (25–7°C) and so produces its worst effects during the summer months when it can complete its life-cycle in a matter of days. Heavily infected animals have many small white pustules on the skin each of which houses a single adult parasite. The dorsal surface and the gills are worst affected. The skin of the lateral regions may be shed and the cornea is often infected, resulting in blindness. Infected fish are difficult to treat as the parasite is under the skin and not easily reached by washes. Prevention is better than cure and a fast water flow combined with low fish density greatly reduces the chances of a serious outbreak. It is important to remove adults from spawning ponds at the earliest opportunity to prevent the vulnerable young fish from becoming infected.

The mobilinid *Trichodina* (Fig. 4B) is a common and widespread ciliate living as an ectoparasite on the skin, gills and mucous membranes of many aquatic animals, both freshwater and marine. As with other ciliate fish parasites, outbreaks in culture ponds are favoured by small ponds with high fish densities and a sluggish water flow. The parasites adversely affect respiratory and ionic exchange at the gills by stimulating mucus production. During infection the number of mucus-producing cells in the gills increases in response to the irritation. Besides infecting fish, *Trichodina* species are found associated with a wide range of other animals. *T. pediculus* is commonly seen on the surface of *Hydra,* while other species are found in the urinary bladder of amphibia and yet others live on the gills and palps of bivalves.

COMMENSALS

The first stomach, or rumen, of ruminants contains large numbers of ciliates (see reviews by Hungate, 1955; 1966). They mainly belong to two orders, the Trichostomatida and the Entodiniomorphida. Of the holotrichs two genera predominate, *Isotricha* and *Dasytricha*. One gymnostome (*Bütschlia*) is occasionally found along with one or two other varieties. However the bulk of the animals are entodiniomorphs including *Entodinium*, *Diplodinium*, *Epidinium*, *Ophryoscolex*, *Caloscolex* and *Opisthotrichum*. The ciliates from sheep and cattle have been investigated in some detail because of their possible nutritional importance to these farm animals. It is likely that the general conclusions hold good for other ruminants.

At birth, calves and lambs are free of ciliates. They become infected from the saliva of adults which is contaminated with ciliates regurgitated during 'cud' chewing. If the young animals are completely isolated from the adults they do not become infected. There is good evidence that the ciliates cannot encyst and do not survive for long outside the host. The ciliates normally appear in lambs and calves within a week of birth but do not become established for some time because of the acidity of the rumen (which results from the fermentation of milk). When less fermentable feed is taken the acidity of the rumen falls and at pH 6·5 the ciliates become numerous. It is possible to inoculate experimentally a young ruminant with a single species of ciliate to produce a large convenient monoculture.

Rumen ciliates are present in huge numbers. Estimates vary from host to host and according to diet, but 1 million/ml is a fairly typical figure (Hungate, 1966). In a cow with a rumen volume of perhaps 100 l the total population of ciliates could well number 10^{11}. All the ciliates are anaerobes and can survive for only a short time in the presence of oxygen. Despite this and their rather specialized way of life, it is possible to culture them *in vitro*. The ruminants' normal food (e.g. grass, dried forage, cereal, etc.) is adequate to support these cultures and there is no requirement for rumen fluid or any other complex organic materials, as was once thought to be the case.

Normally the ciliates live in a protected environment at constant temperature and with a more or less constant supply of food. Do they in their turn benefit the host? Each day a bovine eats about 69 per cent of its ciliates and so it is possible that it could contribute significantly to the nutrition of the host. Hungate (1955) reviews this subject and quotes two estimates for the amount of protein that a bovine might obtain from its ciliates per day. These estimates are in reasonable agreement: between 66–100 g per day, which amounts to some 20 per cent of the animal's daily requirement. It should also be borne in mind that this is animal protein easily utilized and of high nutritional quality. The ciliates also contain large quantities of storage carbohydrate. Is this of significance to the host? It has been calculated that for sheep about 5 g per day of polysaccharide passes to the digestive regions of the gut in this way. This is only about 1 per cent of the daily requirement. However, the ciliates do contribute a significant amount of lactic, acetic and butyric acids to the host. Hungate has estimated that the protozoa provide about 20 per cent of the fermentation products available to the host. Thus it appears these symbionts supply about one-fifth of the ruminants' requirements for organic acids and protein. These conclusions agree remarkably well with the guesses made in 1843 by Graby and Delaford. However, it should be stressed that it is possible to rear perfectly healthy sheep and cattle entirely free of protozoa.

REFERENCES

ADOUTTE, A., and BEISSON, J. (1972). Evolution of mixed populations of genetically different mitochondria in *Paramecium aurelia*. *Nature, Lond.*, **235**, 393–6.

ALLEN, R. D. (1967). Fine structure, reconstruction and possible functions of components of the cortex of *Tetrahymena pyriformis*. *J. Protozool.*, **14**, 553–65.

ALLEN, R. D. (1968). A reinvestigation of cross-sections of cilia. *J. Cell Biol.*, **37**, 825–31.

ALLEN, R. D. (1969). The morphogenesis of basal bodies and accessory structures of the cortex of the ciliated protozoan, *Tetrahymena pyriformis*. *J. Cell Biol.*, **40**, 716–33.

ALLEN, R. D. (1971). Fine structure of membranous and microfibrillar systems in the cortex of *Paramecium caudatum*. *J. Cell Biol.*, **49**, 1–20.

ALLEN, R. D. (1973). Structures linking the myonemes, endoplasmic reticulum and surface membranes in the contractile ciliate *Vorticella*. *J. Cell Biol.*, **56**, 559–79.

ALLEN, S. L., and NANNEY, D. L. (1958). An analysis of nuclear differentiation in the selfers of *Tetrahymena*. *Am. Nat.*, **92**, 139–60.

ALLISON, A. C., HULANDS, G. H., NUNN, J. F., KITCHING, J. A., and MACDONALD, A. C. (1970). The effect of inhalational anaesthetics on the microtubular system in *Actinosphaerium nucleofilum*. *J. Cell Sci.*, **7**, 483–99.

AMMERMAN, D. (1965). Cytologische und genetische Untersuchungen an dem Ciliaten *Stylonychia mytilus* Ehrenberg. *Arch. Protistenk.*, **108**, 109–52.

AMMERMAN, D. (1970). The micronucleus of the ciliate *Stylonuchia mytilus*: its nucleic acid synthesis and its function. *Expl. Cell Res.*, **61**, 6–12.

AMOS, W. B. (1971). A reversible mechanochemical cycle in the contraction of *Vorticella*. *Nature, Lond.*, **299**, 127–8.

AMOS, W. B. (1972). Structure and coiling of the stalk in the peritrich ciliates *Vorticella* and *Carchesium*. *J. Cell Sci.*, **10**, 95–122.

ANDRÉ, J., and FAURÉ-FREMIET, E. (1967). Formation et structures des concrétions calcaires chez *Prorodon morgani* Kahl. *J. Microscopie*, **6**, 391–8.

ANDRUS, W. de W., and GIESE, A. C. (1963). Mechanisms of sodium and potassium regulation in *Tetrahymena pyriformis* Strain W. *J. cell. comp. Physiol.*, **61**, 17–30.

ASTERITA, H., and MARSLAND, D. (1961). The pellicle as a factor in the stabilization of cellular form and integrity: effects of externally applied enzymes on the resistance of *Blepharisma* and *Paramecium* to pressure-induced cytolysis. *J. cell. comp. Physiol.*, **58**, 45–63.

BAKER, H., FRANK, O., PASHER, I., DINNERSTEIN, A., and SOBOTKA, H. (1960). An assay for pantothenic acid in biological fluids. *Clinical Chem.*, **6**, 36–42.

BAKER, J. R. (1969). *Parasitic Protozoa*. Hutchinson, London.

BANNISTER, L. H. (1972). The structure of trichocysts in *Paramecium aurelia*. *J. Cell Sci.* **11**, 899–929.

BANNISTER, L. H., and TATCHELL, E. C. (1968). Contractility and the fibre system of *Stentor coeruleus*. *J. Cell Sci.*, **3**, 295–306.

BAUER, O. N. (1961). Parasitic diseases of cultured fishes and methods of their prevention and treatment. *Parasitology of Fishes*, V. A. Dogiel, G. K. Petrushevski and Y. I. Polyanski (Eds.), 265–98. Oliver and Boyd, London.

BEALE, G. H. (1954). *The Genetics of Paramecium aurelia*. Cambridge University Press, Cambridge.

BEALE, G. H., JURAND, A., and PREER Jr., J. R. (1969). The classes of endo-symbiant of *Paramecium aurelia*. *J. Cell Sci.*, **5**, 65–91.

BEALE, G. H., KNOWLES, J. K. C., and TAIT, A. (1972). Mitochondrial genetics in *Paramecium*. *Nature, Lond.*, **235**, 396–7.

BEISSON, J. and SONNEBORN, T. M. (1965). Cytoplasmic inheritance of the organization of the cell cortex in *Paramecium aurelia*. *Proc. natn. Acad. Sci. U.S.*, **53**, 275–82.

BENDALL, J. R. (1968). *Muscles, Molecules and Movement*. Heineman, London.

BOGGS, N., Jr. (1965). Comparative studies on *Spirostomum*: silver impregnation of three species. *J. Protozool.*, **12**, 603–6.

BOYDE, A., and BARBER, V. C. (1969). Freeze-drying methods for the scanning electron-microscopical study of the protozoan *Spirostomum ambiguum* and the statocyst of the cephalopod mollusc *Loligo vulgaris*. *J. Cell Sci.*, **4**, 223–39.

BOZLER, E. (1924). Uber die Morphologie de Ernährungsorganellen und die Physiologie der Nahnungsanfrahme von *Paramecium caudatum* Ehrb. *Arch. Protistenk.*, **49**, 163–215.

BROKAW, C. J. (1961). Movement and nucleoside polyphosphate activity of isolated flagella from *Polytoma uvella*. *Expl. Cell Res.*, **22**, 151–62.

BRUCE, A. M. (1969). Percolating filters. *Process Biochem.*, **4**, 19–23.

BUTZEL, H. M., Jr., and BOLTEN, A. B. (1968). The relationship of the nutri-tive state of the prey organism, *Paramecium aurelia* to the growth and encystment of *Didinium vestum*. *J. Protozool.*, **15**, 256–8.

BYFIELD, J. E., and LEE, Y. C. (1970). Do synchronizing temperature shifts inhibit RNA synthesis in *Tetrahymena pyriformis*? *J. Protozool.*, **17**, 445–53.

CALKINS, G. N., and SUMMERS, F. M. (Eds.) (1941). *Protozoa in Biological Research*. Columbia University Press, New York.

CAMPBELL, A. S. (1954). Tintinnina. In *Treatise on Invertebrate Paleontology*, Part D, Protista 3, R. C. Moore (Ed.), D166–D180. The Geological Society of America and the University of Kansas, Kansas.

CARASSO, N., FAURÉ-FREMIET, E., and FAVARD, P. (1962) Ultrastructure de l'appareil excréteur chez quelques cilié Péritriches. *J. Microscopie*, **1**, 455–68.

CARASSO, N., and FAVARD, P. (1966). Mise en évidence du calcium dans les myonèmes pedonuclaires de Ciliés Péritriches. *J. Microscopie*, **5**, 759–70.

CARASSO, N., FAVARD, P., and GOLDFISCHER, S. (1964). Localisation à

186 The Ciliates

l'échelle des ultrastructures d'activités de phosphatases en rapport avec les processus digestifs chez un cilié *Campanella umbellaria*. *J. Microscopie*, **3**, 297–322.

CARTER, L. A. (1957). Ionic regulation in the ciliate *Spirostomum ambiguum*. *J. exp. Biol.*, **34**, 71–84.

CHATTON, É., and LWOFF, A. (1935). Les ciliés apostomes. I. Apercu historique et général; étude monographique des genres et des espèces. *Archs. Zool. exp. gen.*, **77**, 1–453.

CHEN, T.-T. (Ed.) (1967a). *Research in Protozoology*, vol. 1. Pergamon Press, Oxford.

CHEN, T.-T. (Ed.) (1967b). *Research in Protozoology*, vol. 2. Pergamon Press, Oxford.

CHEN, T.-T. (Ed.) (1969). *Research in Protozoology*, vol. 3. Pergamon Press, Oxford.

CHEN-SHAN, L. (1969). Cortical morphogenesis in *Paramecium aurelia* following amputation of the posterior region. *J. exp. Zool.*, **170**, 205–28.

CHILD, F. M. (1965). Ciliary co-ordination in glycerinated mussel gills. In *Progress in Protozoology*, 11, Internat. Cong. Ser. **91**, Excerpta Medica Foundation, London.

CHILD, F. M. (1967). The chemistry of protozoan cilia and flagella. In *Chemical Zoology*, vol. 1, *Protozoa*, M. Florkin, B. T. Scheer and G. W. Kidder (Eds.), 381–93. Academic Press, London.

CLEVELAND, L. R., and GRIMSTONE, A. V. (1964). The fine structure of the flagellate *Mixotricha paradoxa* and its associated micro-organisms. *Proc. R. Soc. B*, **59**, 668–86.

CONNOR, R. L. (1967). Transport phenomena in Protozoa. In *Chemical Zoology*, vol. 1, *Protozoa*, M. Florkin, B. T. Scheer and G. W. Kidder (Eds.), 309–50. Academic Press, London.

CORLISS, J. O. (1953). Silver impregnation of ciliated protozoa by the Chatton–Lwoff technique. *Stain Technol.*, **28**, 97–100.

CORLISS, J. O. (1955). The opalinid infusorians: flagellates or ciliates? *J. Protozool.*, **2**, 107–14.

CORLISS, J. O. (1956). On the evolution and systematics of ciliated Protozoa. *Syst. Zool.*, **5**, 68–91; 121–40.

CORLISS, J. O. (1958a). The systematic position of *Pseudomicrothorax dubius*, ciliate with a unique combination of anatomical features. *J. Protozool.*, **5**, 184–93.

CORLISS, J. O. (1958b). The phylogenetic significance of the genus *Pseudomicrothorax* in the evolution of holotrichous ciliates. *Acta biol. Acad. Sci. hung.*, **8**, 367–88.

CORLISS, J. O. (1959). An illustrated key to the higher groups of the ciliated Protozoa, with definition of terms. *J. Protozool.*, **6**, 265–81.

CORLISS, J. O. (1961). *The Ciliated Protozoa*. Pergamon Press, Oxford.

CURDS, C. R. (1963). The flocculation of suspended matter by *Paramecium caudatum*. *J. gen. Microbiol.*, **33**, 357–63.

CURDS, C. R. (1966). An ecological study of the ciliated protozoa in activated sludge. *Oikos*, **15**, 282–9.

CURDS, C. R. (1969). *An Illustrated Key to the British Freshwater Ciliated Protozoa Commonly Found in Activated Sludge*. Water Pollution Research Technical Paper No. 12. HMSO, London.

CURDS, C. R., and COCKBURN, A. (1968). Studies on the growth and feeding of *Tetrahymena pyriformis* in axenic and monoxenic culture. *J. gen. Microbiol.*, **54**, 343–58.

CURDS, C. R., and COCKBURN, A. (1970). Protozoa in biological sewage-treatment processes. II. Protozoa as indicators in the activated sludge process. *Water Res.*, **4**, 237–49.

CZARSKA, L. (1964). Role of K $^+$ and Ca $^{2+}$ ions in the exitability of protozoan cell. Chemical and electric stimulation of contractile vacuoles. *Acta Protozool.*, **2**, 287–96.

DANFORTH, W. F. (1967). Respiratory metabolism. In *Research in Protozoology*, vol. 1, T.-T. Chen (Ed.), 201–306. Pergamon Press, Oxford.

DANIELLI, J. F. (1959). Theoretical aspects of nucleo-cytoplasmic relationships. *Expl. Cell Res.*, **6** (*suppl.*), 252–67.

DE TERRA, N. (1960). A study of nucleo-cytoplasmic interactions during cell division in *Stentor coeruleus*. *Expl. Cell. Res*, **21**, 41–8.

DE TERRA, N. (1966). Culture of *Stentor coeruleus* on *Colpidium campylum*. *J. Protozool.*, **13**, 491–2.

DE TERRA, N. (1970). Cytoplasmic control of macronuclear events in the cell cycle of *Stentor*. *Symp. Soc. exp. Biol.*, **24**, 345–68.

DEWEY, V. C. (1967). Lipid composition, nutrition and metabolism. In *Chemical Zoology*, vol. 1, *Protozoa*, M. Florkin, B. T. Scheer and G. W. Kidder (Eds.), 162–274. Academic Press, London.

DIPPELL, R. V. (1962). The site of silver impregnation in *Paramecium aurelia*. *J. Protozool.*, **9** (*suppl.*), 24.

DIPPELL, R. V. (1968). The development of basal bodies in *Paramecium*. *Proc. natn. Acad. Sci. U.S.*, **61**, 461–8.

DOBELL, C. C. (1932). *Anton van Leeuwenhoek and his little animals*. Bate, Son and Danielsson, London.

DOGIEL, V. A. (1965). *General Protozoology*. Clarendon Press, Oxford.

DRAGESCO, J. (1952). Sur la structure des trichocystes toxiques des infusoires holotriches gymnostomes. *Bull. Microsc. appl.*, **2**, 92–8.

DRAGESCO, J. (1962). Capture et ingestion des proies chez les infusoires ciliés. *Bull. Biol. Fr. Belg.*, **96**, 123–67.

DRYL, S. (1970). Response of ciliate Protozoa to external stimuli. *Acta Protozool.*, **7**, 325–33.

DUNHAM, P. B. (1969). Regulation of intracellular sodium in *Tetrahymena*. In *Progress in Protozoology*, A. A. Strelkov, K. M. Sukhanova and I. B. Raikov (Eds.), 138. Nauka, Leningrad.

DUNHAM, P. B., and CHILD, F. M. (1961). Ion regulation in *Tetrahymena*.. *Biol. Bull.*, **121**, 129–40.

DUNHAM, P. B., and STONER, L. C. (1969). Indentation of the pellicle of *Tetrahymena* at the contractile vacuole pore before systole. *J. Cell Biol.*, **43**, 184–8.

ECKERT, R., and NAITOH, Y. (1970). Passive electrical properties of *Paramecium* and problems of ciliary coordination. *J. gen. Physiol.*, **55**, 467–83.

EHRET, C. F., and POWERS, E. L. (1959). The cell surface of *Paramecium*. *Int. Rev. Cytol.*, **8**, 97–133.

ELLIOTT, A. M., and BAK, I. J. (1964a). The contractile vacuole and related structures in *Tetrahymena pyriformis*. *J. Protozool.*, **11**, 250–61.

ELLIOTT, A. M., and BAK, I. J. (1964b). The fate of mitochondria during ageing in *Tetrahymena pyriformis*. *J. Cell Biol.*, **20**, 113–29.

ELLIOTT, A. M., and CLEMMONS, G. L. (1966). An ultrastructural study of ingestion and digestion in *Tetrahymena pyriformis*. *J. Protozool.*, **13**, 311–23.

ELLIOTT, A. M., and ZIEG, R. G. (1968). A Golgi apparatus associated with mating in *Tetrahymena pyriformis*. *J. Cell Biol.*, **36**, 391–8.

ESTÉVÉ, J.-C. (1970). Distribution of acid phosphatase in *Paramecium caudatum:* its relations with the process of digestion. *J. Protozool.,* **17,** 24–35.

ESTÉVÉ, J.-C. (1972). L'appareil de Golgi des ciliés. Ultrastructure, particulierement chez *Paramecium. J. Protozool.,* **19,** 609–18.

ETTIENNE, E. M. (1970). Control of contractility in *Spirostomum* by dissociated calcium ions. *J. gen. Physiol.,* **56,** 168–79.

FALK, H., WUNDERICH, F., and FRANKE, W. W. (1968). Microtubular structures in macronuclei of synchronously dividing *Tetrahymena pyriformis. J. Protozool.,* **15,** 776–80.

FAURÉ-FREMIET, E. (1924). Contribution à la connaissance des infusoires planktoniques. *Bull. biol. Fr. Belg.,* suppl. no. **6,** 1–171.

FAURÉ-FREMIET, E. (1950). Morphologie comparée et systématique des ciliés. *Bull. Soc. Zool. Fr.,* **75,** 109–22.

FAURÉ-FREMIET, E. (1954). Réorganisation du type endomixique chez les Loxodidae et chez les *Centrophorella. J. Protozool.,* **1,** 20–7.

FAURÉ-FREMIET, E. (1957). Le macronucleus hétéromere de quelques Ciliés. *J, Protozool.,* **4,** 7–17.

FAURÉ-FREMIET, E. (1961). Remarques sur la morphologie comparée et la systématique des Ciliata Hypotrichida. *C. R. Acad. Sci. Paris,* **252,** 3515–19.

FAURÉ-FREMIET, E., ANDRÉ, J., and GANIER, M. C. (1968). Calcification tegumentaire chez les ciliés du genre *Coleps* Nitzsch. *J. Microscopie,* **7,** 693–704.

FAURÉ-FREMIET, E., FAVARD, P., and CARASSO, N. (1962). Étude au microscope électronique des ultrastructures d'*Epistylis anastatica* (Vilié peritriche). *J. Microscopie,* **1,** 287–312.

FAURÉ-FREMIET, E., ROUILLER, C., and GAUCHERY, M. (1957). La réorganisation macronucléaire chez les *Euplotes.* Étude au microscope électronique. *Expl. Cell Res.,* **12,** 135–44.

FINLEY, H. E., BROWN, C. A., and DANIEL, W. A. (1964). Electron microscopy of the ectoplasm and infraciliature of *Spirostomum ambiguum. J. Protozool.,* **11,** 264–80.

FLAVIN, M., and GRAFF, S. (1951). Utilization of guanine for nucleic acid biosynthesis by *Tetrahymena geleii. J. biol. Chem.,* **191,** 55–61.

FLORKIN, M., SCHEER, B. T., and KIDDER, G. W. (Eds.) (1967). *Chemical Zoology,* vol. 1, *Protozoa.* Academic Press, London.

FURGASON, W. H. (1940). The significant cystostomal pattern of the "*Glacoma-Colpidium* group", and a proposed new genus and species, *Tetrahymena geleii. Arch. Protistenk.,* **67,** 376–500.

GALL, J. G. (1959). Macronuclear duplication in the ciliated protozoan *Euplotes. J. biophys. biochem. Cytol.,* **5,** 295–307.

GIBBONS, I. R. (1963). Studies on the protein components of cilia from *Tetrahymena pyriformis. Proc. natn. Acad. Sci. U.S.,* **50,** 1002–10.

GIBBONS, I. R. (1965). Chemical dissection of cilia. *Archs. Biol., Paris,* **76,** 317–52.

GIBBONS, I. R. (1966). Studies on the adenosine triphosphatase activity of 14S and 30S dynein from cilia of *Tetrahymena. J. biol. Chem.,* **241,** 5590–6.

GIBBONS, I. R., and GRIMSTONE, A. V. (1960). On flagellar structure in certain flagellates. *J. biophys. biochem. Cytol.,* **7,** 697–716.

GIBBONS, I. R., and ROWE, A. J. (1965). Dynein: a protein with adenosine triphosphatase activity from cilia. *Science,* **149,** 424–6.

GLIDDON, R. (1965). Ciliary activity and coordination in *Euplotes eurystomus.* In *Progress in Protozoology,* Internat. Congr. Ser. 91, Excerpta Medica Foundation, London, 216.

GLIDDON, R. (1966). Ciliary organelles and associated fibre systems in *Euplotes eurystomus* (Ciliata, Hypotrichida). I. Fine structure. *J. Cell Sci.,* **1,** 439–48.

GOLDSTEIN, L. (1963). RNA and protein in nucleocytoplasmic interactions. In *Cell Growth and Cell Division,* vol. 2, R. J. C. Harris (Ed.), 129–49. Academic Press, London.

GRAIN, J. (1968). Les systèmes fibrillaires chez *Stentor igneus* Ehrenberg et *Spirostomum ambiguum* Ehrenberg. *Protistologica,* **4,** 27–35.

GRAIN, J. (1969). Le cinétosome et ses dérivés chez les ciliés. *Ann. Biol.,* **8,** 53–97.

GRAY, J. (1930). The mechanism of ciliary movement. VI. Photographic and stroboscopic analysis of ciliary movement. *Proc. R. Soc. B.,* **107,** 313–32.

GREBECKI, A. (1965). Role of Ca^{2+} ions in the excitability of protozoan cell. Decalcification, recalcification, and the ciliary reversals in *Paramecium caudatum. Acta Protozool.,* **3,** 275–90.

GRELL, K. G. (1962). Morphologie und Fortpflangung der Protozoen. (Einschliesslich Entwichlungsphysiologie und Genetik.) *Fortschr. Zool.,* **14,** 1–85.

GRELL, K. G. (1964). The protozoan nucleus. In *The Cell,* vol. 6, J. Brachet and A. Mirsky (Eds.), 1–79. Academic Press, London.

GRELL, K. G. (1967). Sexual reproduction in Protozoa. In *Research in Protozoology,* vol. 2, T.-T. Chen (Ed.), 147–213. Pergamon Press, Oxford.

GRELL, K. G. (1968). *Protozoologie.* Springer-Verlag, Wien.

GRIMSTONE, A. V. (1961). Fine structure and morphogenesis in Protozoa. *Biol. Rev.,* **36,** 97–150.

GRIMSTONE, A. V., and KLUG, A. (1966). Observations on the substructure of flagellar fibres. *J. Cell Sci.,* **1,** 351–62.

GUILCHER, Y. (1948). Affinités structurales des bourgeons migrateurs d'infusoires acinétiens. *C.R. Acad. Sci., Paris,* **17,** 1304–8.

GUILCHER, Y. (1951). Contribution à l'étude des ciliés gemmipores, chonotriches et tentaculifères. *Ann. Sci. nat., Zool. (sér.* 11), **13,** 33–132.

GURDON, J. B. (1970). The autonomy of nuclear activity in multicellular organisms. *Symp. Soc. exp. Biol.,* **24,** 369–78.

GUTTES, E., and GUTTES, S. (1960). Incorporation of tritium-labelled thymidine into the macronucleus of *Stentor coeruleus. Exp. Cell Res.,* **19,** 626–8.

HAIRSTON, N. G. (1958). Observations on the ecology of *Paramecium* with comments on the species problem. *Evolution,* **12,** 440–50.

HALL, R. P. (1953). *Protozoology.* Prentice Hall, New York.

HALL, R. P. (1965). *Protozoan Nutrition.* Blaisdell, New York.

HARGITT, G. T., and FRAY, W. (1917). The growth of *Paramoecium* in pure cultures of bacteria. *J. exp. Zool.,* **22,** 421–56.

HARRIS, J. E. (1961). The mechanics of ciliary movement. In *The Cell and the Organism,* J. A. Ramsay and V. B. Wigglesworth (Eds.), 22–36. Cambridge University Press, Cambridge.

HILDEN, S. (1970). Sodium and potassium levels in *Blepharisma intermedium. Expl. Cell Res.,* **61,** 246–54.

HMSO (1968). Protozoa in sewage-treatment processes. *Notes on Water Pollution* (Ministry of Technology), no. 43.

HOFFMAN-BERLING, H. (1958). Der mechanismus eines neuen, von der muskelkontraktion verschiedenen Kontraktionszyklus. *Biochim. biophys. Acta,* **27,** 247–55.

HOLZ, G. G. (1964). Nutrition and metabolism of ciliates. In *Biochemistry and Physiology of Protozoa,* vol. 3, S. H. Hutner (Ed.), 199–242. Academic Press, New York.

HOLZ, G. G., Jr., ERWIN, J., WAGNER, B., and ROSENBAUM, N. (1962) The nutrition of *Tetrahymena setifera* H2-1; sterol and alcohol requirements. *J. Protozool.,* **9**, 359–63.

HONIGBERG, B. M., BALAMUTH, W., BOVEE, E. C., CORLISS, J. O., GOJDICS, M., HALL, R. P., KUDO, R. R., LEVINE, N. D., LOEBLICH, A. R., Jr., WEISER, J., and WENRICH, D. H. (1964). A revised classification of the phylum Protozoa. *J. Protozool.,* **11**, 7–20.

HOPKINS, J. M. (1970). Subsidiary components of the flagella of *Chlamydomonas reinhardii. J. Cell. Sci.,* **7**, 823–39.

HULL, R. W. (1961a). Studies on suctorian Protozoa: the mechanism of prey adherence. *J. Protozool.,* **8**, 343–50.

HULL, R. W. (1961b). Studies on suctorian Protozoa: the mechanism of ingestion of prey cytoplasm. *J. Protozool.,* **8**, 351–9.

HUNGATE, R. E. (1955). Mutualistic intestinal protozoa. In *Biochemistry and Physiology of Protozoa,* vol. 2, S. H. Hutner and A. Lwoff (Eds.), 159–99. Academic Press, New York.

HUNGATE, R. E. (1966). *The Rumen and its Microbes.* Academic Press, New York.

HUTNER, S. H. (Ed.) (1964). *Biochemistry and Physiology of Protozoa,* vol. 3. Academic Press, New York.

HUTNER, S. H., and LWOFF, A. (Eds.) (1955). *Biochemistry and Physiology of Protozoa,* vol. 2. Academic Press, New York.

HUXLEY, J. S. (1926). *Essays in Popular Science.* Chatto and Windus, London.

HYMAN, L. H. (1940). *The Invertebrates: Protozoa through Ctenophora.* McGraw-Hill, New York.

JAHN, T. L. (1961). The mechanism of ciliary movement. I. Ciliary reversal and activation by electric current; the Ludloff phenomenon in terms of core and volume conductors. *J. Protozool.,* **8**, 369–80.

JAHN, T. L. (1962). The mechanism of ciliary movement. II. Ion antagonism and ciliary reversal. *J. cell. comp. Physiol.,* **60**, 217–28.

JAHN, T. L., and BOVEE, E. C. (1964). Protoplasmic movements and locomotion of Protozoa. In *Biochemistry and Physiology of Protozoa,* vol. 3, S. H. Hutner (Ed.), 61–129. Academic Press, New York.

JONES, A. R. (1966). Uptake of [45] Ca by *Spirostomum ambiguum. J. Protozool.,* **13**, 422–8.

JONES, A. R. (1967). Calcium and phosphorus accumulation in *Spirostomum ambiguam. J. Protozool.,* **14**, 220–5.

JONES, A. R. (1969). Mitochondria, calcification and waste disposal. *Calc. Tiss. Res.,* **3**, 363–5.

JONES, A. R., JAHN, T. L., and FONSECA J. R. (1966). Contraction of protoplasm. I. Cinematographic analysis of the anodally stimulated contraction of *Spirostomum ambiguum. J. Cell. Physiol.,* **68**, 127–33.

JONES, A. R., JAHN, T. L., and FONESCA, J. R. (1970a). Contraction of protoplasm. III. Cinematographic analysis of the contraction of some heterotrichs. *J. Cell. Physiol.,* **75**, 1–8.

JONES, A. R., JAHN, T. L., and FONESCA, J. R. (1970b). Contraction of protoplasm. IV. Cinematographic analysis of the contraction of some peritrichs. *J. Cell. Physiol.,* **75**, 9–20.

JONES, A. R., and MORLEY, N. (1969). Relaxation of the stalk of *Vorticella. J. Protozool.* (suppl.), **16**, 21.

JURAND, A., BEALE, G. H.,and YOUNG, M. R. (1962). Studies on the macronucleus of *Paramecium aurelia.* I. (With a note on ultraviolet micrography.) *J. Protozool.,* **9**, 122–31.

KAMADA, T. (1940). Ciliary reversal in *Paramecium*. *Proc. Imp. Acad. Tokyo*, **16**, 241–7.

KAMADA, T., and KINOSITA, H. (1940). Calcium–potassium factor in ciliary reversal of *Paramecium*. *Proc. Imp. Acad. Tokyo*, **16**, 125–30.

KANEDA, M. (1962). Fine structure of the oral apparatus of the gymnostome ciliate *Chlamydodon pedarius*. *J. Protozool.*, **9**, 188–95.

KANESHIRO, E. S., DUNHAM, P. B., and HOLZ, G. G. (1969a). Osmoregulation in a marine ciliate, *Miamiensis avidus*. I. Regulation of inorganic ions and water. *Biol. Bull.*, **136**, 63–75.

KANESHIRO, E. S., HOLZ, G. G., and DUNHAM, P. B. (1969b). Osmoregulation in a marine ciliate, *Miamiensis avidus*. II. Regulation of intracellular free aminoacids. *Biol. Bull.*, **137**, 161–9.

KARAKASHIAN, S. (1963). Growth of *Paramecium bursaria* as influenced by the presence of algae symbionts. *Physiol. Zool.*, **36**, 52–67.

KARAKASHIAN, S. J., KARAKASHIAN, M. W., and RUDZINSKA, M. A. (1968). Electron microscopic observations on the symbiosis of *Paramecium bursaria* and its intracellular algae. *J. Protozool.*, **15**, 113–28.

KARAKASHIAN, S. J., and SEIGEL, R. W. (1965). A genetic approach to endocellular symbiosis. *Expl. Parasit.*, **17**, 103–22.

KEHLENBECK, E. K., DUNHAM, P. B., and HOLZ, G. G. (1965). Inorganic ion concentrations in a marine ciliate, *Uronema*. In *Progress in Protozoology*, 245. Excerpta Medica Foundation, Amsterdam.

KENNEDY, J. R., Jr. (1965). The morphology of *Blepharisma undulans* Stein. *J. Protozool.*, **12**, 542–61.

KENT, W S. (1880–2). *A Manual of the Infusoria*, vols. 1–3. David Bogue, London.

KIDDER, G. W. (1967). Nitrogen: distribution, nutrition and metabolism. In *Chemical Zoology*, vol. 1, *Protozoa*, M. Florkin, B. T. Scheer and G. W. Kidder (Eds.), 93–159. Academic Press, London.

KIDDER, G. W., and DEWEY, V. C. (1951). The biochemistry of ciliates in pure culture. In *Biochemistry and Physiology of Protozoa*, vol. 1, A. Lwoff (Ed.), 324–400. Academic Press, New York.

KIMBALL, R. F., and PERDUE, S. W. (1962). Quantitative cytochemical studies of nucleic acid synthesis. *Expl. Cell Res.*, **27**, 405–15.

KIMBALL, R. F., VOGT-KÖHNE, L., and CASPERSON, T. O. (1960). Quantitative cytochemical studies on *Paramecium aurelia*. III. Dry weight and ultraviolet absorption of isolated macronuclei during various stages of the interdivisional interval. *Expl. Cell Res.*, **20**, 368–77.

KING, R. L. (1933). Contractile vacuole of *Euplotes*. *Trans. Am. Microsc. Soc.*, **52**, 103–6.

KINOSITA, H. (1954). Electric potentials and ciliary response in *Opalina*. *J. Fac. Sci. Tokyo Univ.* (Sect. 4), **7**, 1–14.

KINOSITA, H., DRYL, S., and NAITOH, Y. (1964). Spontaneous change in membrane potential of *Paramecium caudatum* induced by calcium and barium ions. *Bull. Acad. polon. Sci. Sér. Sci. biol.*, **12**, 459–61.

KINOSITA, H., and KAMADA, T. (1939). Movement of abfrontal cilia of *Mytilus*. *Jap. J. Zool.*, **8**, 291–310.

KINOSITA, H., and MURIHAMI, A. (1967). Control of ciliary motion. *Physiol. Rev.*, **47**, 53–82.

KITCHING, J. A. (1934). The physiology of contractile vacuoles. I, Osmotic relations. *J. exp. Biol.*, **11**, 364–81.

KITCHING, J. A. (1936). The physiology of contractile vacuoles. II. The control of body volume in marine peritrichs. *J. exp. Biol.*, **13**, 11–27.

KITCHING, J. A. (1938). Contractile vacuoles. *Biol. Rev.*, **13**, 403–44.

KITCHING, J. A. (1948a). Physiology of contractile vacuoles. V. The effects of short-term variations of temperature on a freshwater peritrich ciliate. *J. exp. Biol.*, **25**, 406–20.

KITCHING, J. A. (1948b). The physiology of contractile vacuoles. VI. Temperature and osmotic stress. *J. exp. Biol.*, **25**, 421–36.

KITCHING, J. A. (1952a). Observations on the mechanism of feeding in the suctorian *Podophyra*. *J. exp. Biol.*, **29**, 255–66.

KITCHING, J. A. (1952b). The physiology of contractile vacuoles. VIII. The water relations of the suctorian *Podophrya* during feeding. *J. exp. Biol.*, **29**, 363–71.

KITCHING, J. A. (1954a). The physiology of contractile vacuoles. IX. Effects of sudden changes in temperature on the contractile vacuole of a suctorian; with a discussion of the mechanism of contraction. *J. exp. Biol.*, **31**, 68–75.

KITCHING, J. A. (1954b). The physiology of contractile vacuoles. X. Effects of high hydrostatic pressure on the contractile vacuole of a suctorian. *J. exp. Biol.*, **31**, 76–83.

KITCHING, J A. (1956a). Contractile vacuoles of Protozoa. *Protoplasmotologia*, **III D3a**, 1–45.

KITCHING, J. A. (1956b). Food vacuoles. *Protoplasmatologia*, **III D3b**, 1–54.

KITCHING, J. A. (1967). Contractile vacuoles, ionic regulation and excretion. In *Research in Protozoology*, vol. 1, T.-T. Chen (Ed.) 307–36. Pergamon Press, Oxford.

KLEIN, R. L. (1961). Homeostatic mechanisms for cation regulation in *Acanthamoeba* sp. *Expl. Cell Res.*, **25**, 571–84.

KOFOID, C. A. (1930). Factors in the evolution of the pelagic ciliate, the Tintinoinea. In *Contributions to Marine Biology*, 1–39. Stanford University Press, California.

KOZLOFF, E. N. (1954). Studies on an astomatous ciliate from a freshwater limpet, *Ferrissia peninsulae*. *J. Protozool.*, **1**, 200–6.

KUDO, R. R. (1954). *Protozoology*. Thomas, Springfield, Illinois.

KUZNICKI, L., JAHN, T. L., and FONSECA, J. R. (1970). Helical nature of the ciliary beat of *Paramecium multimicronucleatum*. *J. Protozool.*, **17**, 16–24.

LEGRAND, B. (1968). Essai de fixation en extension du spirostome après anesthésie préalable. *Protistologica*, **4**, 263–70.

LEVINE, L. (1956). Contractility of glycerinated vorticellae. *Biol. Bull.*, **111**, 319.

LILLY, D. M. (1967). Growth factors in Protozoa. In *Chemical Zoology*, vol. 1, *Protozoa*, M. Florkin, B. T. Scheer and G. W. Kidder (Eds.), 275–307. Academic Press, London.

LOEFER, J. B. (1936). Bacteria-free culture of *Paramecium bursaria* and concentration of the medium as a factor in growth. *J. exp. Zool.*, **72**, 387–407.

LOEFER, J. B., SMALL, E. B., and FURGASON, W. H. (1966). Range and variation in somatic infraciliature and contractile vacuole pores of *Tetrahymena pyriformis*. *J. Protozool.*, **13**, 90–102.

LOM, J., and KOZLOFF, E. N. (1967). The ultrastructure of *Phalacrocleptes verruciformis*, an unciliated ciliate parasitising the polychaete *Schizobranchia insignis*. *J. Cell. Biol*, **33**, 355–64.

LOWNDES, A. G. (1943). The swimming of unicellular flagellate organisms. *Proc. zool. Soc. Lond. A*, **113**, 99–107.

LUDWIG W. (1928), Der Betriebstoffwechsel von *Paramecium caudatum* Ehrbg. *Arch. Protistenk.*, **62**, 12–40.

LWOFF, A. (1950). *Problems of Morphogenesis in Ciliates*. Wiley and Sons, New York.

LWOFF, A. (Ed.) (1951). *Biochemistry and Physiology of Protozoa*, vol. 1. Academic Press, New York.

MCDONALD, B. B. (1962). Synthesis of deoxyribonucleic acid by micro- and macronuclei of *Tetrahymena pyriformis*. *J. Cell Biol.*, **13**, 193–203.

MACKINNON, D. L., and HAWES, R. S. J. (1961). *An Introduction to the Study of Protozoa*. Clarendon Press, Oxford.

MCLAUGHLIN, J. A., and ZAHL, P. (1966). Endozoic algae. In *Symbiosis*, vol. 1, S. Henry (Ed.), 57–97. Academic Press, New York.

MACLENNAN, R. F. (1944). The pulsatory cycle of the contractile canal in the ciliate *Hatophyra*. *Trans. Am. microsc. Soc.*, **63**, 187–98.

MANDEL, M. (1967). Nucleic acids of Protozoa. In *Chemical Zoology*, vol.1, *Protozoa*, M. Florkin, B. T. Scheer and G. W. Kidder (Eds.), 541–72. Academic Press, London.

MARGULIS, L. (1971). Symbiosis and evolution. *Sci. Amer.*, **225**, 48–57.

MAYR, E. (1957). Species concepts and definitions. In *The Species Problem*, E. Mayr (Ed.), 1–22. Amer. Assoc. Adv. Sci., Washington.

MILLECCHIA, L., and RUDZINSKA, M. A. (1968). An abnormality in the life-cycle of *Tokophrya infusionum*. *J. Protozool.*, **15**, 665–73.

MILLER, S. (1968). The predatory behaviour of *Dileptus anser*. *J. Protozool.*, **15**, 313–19.

MÜLLER, M. (1967). Digestion. In *Chemical Zoology*, vol. 1, *Protozoa*, M. Florkin, B. T. Scheer and G. W. Kidder (Eds.), 351–80. Academic Press, London.

MÜLLER, M., RÖHLICH, P., and TÖRÖ, I. (1965). Studies on feeding and digestion in Protozoa. VII. Ingestion of polystyrene latex particles and its early effect on acid phosphatase in *Paramecium multimicronucleatum* and *Tetrahymena pyriformis*. *J. Protozool.*, **12**, 27–34.

MÜLLER, M., RÖHLICH, P., TÓTH, J., and TÖRÖ, I. (1963). Fine structure and enzymic activity of protozoan food vacuoles. In *Ciba Foundation Symposium on Lysosomes*, 201–16. Churchill, London.

MURTI, K. G., and PRESCOTT, D. M. (1970). Micronuclear ribonucleic acid in *Tetrahymena pyriformis*. *J. Cell. Biol.*, **47**, 460–67.

MUSCATINE, L., KARAKASHIAN, S. J., and KARAKASHIAN, M. W. (1967). Soluble extracellular products of algae symbiotic with a ciliate, a sponge and a mutant hydra. *Comp. Biochem. Physiol.*, **20**, 1–12.

NAITOH, Y. (1968). Ionic control of the reversal response of cilia in *Paramecium caudatum*. A calcium hypothesis. *J. gen. Physiol.*, **51**, 85–103.

NAITOH, Y., and ECKERT, R. (1969a). Ionic mechanisms controlling behavioural responses of *Paramecium* to mechanical stimulation. *Science*, **164**, 963–5.

NAITOH, Y., and ECKERT, R. (1969b). Ciliary orientation: controlled by cell membrane or by intracellular fibrils? *Science*, **166**, 1633–5.

NANNEY, D. L., and RUDZINSKA, M. A. (1960). Protozoa. In *The Cell*, vol. 4, J. Brachet and A. E. Mirsky (Eds.), 109–50. Academic Press, New York.

NEWTON, B. A. (1957). Nutritional requirements and biosynthetic capabilities of the parasitic flagellate *Strigomonas oncopelti*. *J. gen. Microbiol.*, **17**, 708–17.

NILSSON, J. R. (1970). Suggestive structural evidence for macronuclear 'subnuclei' in *Tetrahymena pyriformis* GL. *J. Protozool.*, **17**, 539–48.

NOIROT-TIMOTHÉE, C. (1960). Étude d'une famille de ciliés: les Ophyroscolecidae. Structures et ultrastructures. *Ann. Sci. nat. (Zool.)*, **2**, 527–718.

G

NOLAND, L. E., and FINLEY, H. E. (1931). Studies on the taxonomy of the genus *Vorticella*. *Trans. Am. Microsc. Soc.*, **50**, 81–123.

NOVIKOFF, A. B. (1961). Lysosomes and related particles. In, *The Cell*, vol. 2, J. Brachet and A. Mirsky (Eds.), 423–88. Academic Press, New York.

OKIJIMA, A., and KINOSITA, H. (1966). Ciliary activity and coordination in *Euplotes eurystomus*. I. Effect of microdissection of neuromotor fibres. *Comp. Biochem. Physiol.*, **19**, 115–31.

ORGAN, A. E., BOVEE, E. C., and JAHN, T. L. (1968a). Adenosine triphosphate acceleration of the nephridial apparatus of *Paramecium multimicronucleatium*. *J. Protozool.*, **15**, 173–6.

ORGAN, A. V., BOVEE, E. C., JAHN, T. L., WIGG, D., and FONESCA, J. R. (1968b). The mechanism of the nephridial apparatus of *Paramecium multinucleatum*. I. Expulsion of water from the vesicle . *J. Cell Biol.*, **37**, 139–45.

ORGAN, A. E., BOVEE, E. C., and JAHN, T. L. (1972). The mechanism of the water expulsion vesicle of the ciliate *Tetrahymena pyriformis*. *J. Cell Biol.*, **55**, 644–52.

PÁRDUCZ, B. (1966). Ciliary movement and coordination in ciliates. *Int. Rev. Cytol.*, **21**, 91–128.

PAUTARD, F. G. E. (1959). Hydroxyapatite as a developmental feature of *Spirostomum ambiguum*. *Biochim. biophys. Acta.*, **35**, 33–46.

PAUTARD, F. G. E. (1962). Biomolecular aspects of spermatozoan motility. In *Spermatozoan Motility*, D. W. Bishop (Ed.), 189–232. Am. Assoc. Adv. Sci., Washington, D.C.

PAUTARD, F. G. E. (1970). Calcification in unicellular organisms. In *Biological Calcification: Cellular and Molecular Aspects*, H. Schraer (Ed.), 105–202. North-Holland, Amsterdam.

PHILPOT, C. H. (1928). Growth of *Paramoecia* in pure cultures of pathogenic bacteria and in the presence of soluble products of such bacteria. *J. Morph.*, **46**, 85–129.

PICKEN, L. E. R. (1936). A note on the mechanism of salt and water balance in the heterotrichous ciliate, *Spirostomum ambiguum*. *J. exp. Biol.*, **13**, 387–92.

PITELKA, D. R. (1961). Fine structure of the silverline and fibrillar systems of three tetrahymenid ciliates. *J. Protozool.*, **8**, 75–89.

PITELKA, D. R. (1963). *Electron-microscopic Structure of Protozoa*. Pergamon Press, Oxford.

PITELKA, D. R. (1965). New observations on cortical ultrastructure in *Paramecium*. *J. Microscopie*, **4**, 373–94.

PITELKA, D. R. (1969). Fibrillar systems in Protozoa. In *Research in Protozoology*, vol. 3, T.-T. Chen (Ed.), 279–388. Pergamon Press, Oxford.

PITELKA, D. R. (1970). Ciliate ultrastructure: some problems in cell biology. *J. Protozool.*, **17**, 1–10.

PREER, J. R., Jr. (1969). Genetics of the Protozoa. In *Research in Protozoology*, vol. 3, T.-T. Chen (Ed.), 129–278. Pergamon Press, Oxford.

PREER, L. B., JURAND, A., PREER Jr., J. R., and RUDMAN, B. M. (1972). The classes of kappa in *Paramecium aurelia*. *J. Cell Sci.*, **11**, 581–600.

PRESCOTT, D. M. (1960). Relation between cell growth and cell division. IV. The synthesis of DNA, RNA and protein from division to division in *Tetrahymena*. *Expl. Cell Res.*, **19**, 228–38.

PRESCOTT, D. M., KIMBALL, R. F., and CARRIER, R. F. (1962). Comparison between the timing of micronuclear and macronuclear DNA synthesis in *Euplotes eurystomus*. *J. Cell Biol.*, **13**, 175–6.

PRESCOTT, D. M., and STONE, G. E. (1967). Replication and function of the protozoan nucleus. In *Research in Protozoology*, vol. 3, T.-T. Chen (Ed.), 117–46. Pergamon Press, Oxford.

PRINGSHEIM, E. G. (1928). Physiologische Untersuchungen an *Paramoecium bursaria*. Ein Beitrag zur Symbioseforschung. *Arch. Prostistenk.*, **64**, 291–418.

PRUSCH, R. D., and DUNHAM, P. B. (1970). Contraction of isolated contractile vacuoles from *Amoeba proteus*. *J. Cell Biol.*, **46**, 431–4.

PUYTORAC, P. DE (1954). Contribution à l'étude cytologique et taxonomique des infusoires astomes. *Ann. Sci. nat., Zool.* (ser. 2), **16**, 85–270.

PUYTORAC, P. DE (1959). Quelques observations sur l'évolution et les origines des ciliés Astomatida. *Proc. XV int. Cong. Zool., London*, 649–51.

RAIKOV, I. B. (1958). Der Formwechsel des Kernapparates einiger neiderer Ciliaten. I. Die Gattung *Trachelocerca*. *Arch. Protistenk.*, **130**, 129–92.

RAIKOV, I. B. (1969). The macronucleus of ciliates. In *Research in Protozoology*, vol. 3, T.-T. Chen (Ed.), 1–128. Pergamon Press, Oxford.

RANDALL, J. T. (1957). The fine structure of the protozoan *Spirostomum ambiguum*. *Symp. Soc. exp. Biol.*, **10**, 185–98.

RANDALL, J. T., and HOPKINS, J. M. (1962). On the stalks of certain peritrichs. *Phil. Trans. R. Soc. Ser. B.* **245**, 59–79.

RANDALL, J. T., and JACKSON, S. F. (1958). Fine structure and function in *Stentor polymorphus*. *J. biophys. biochem. Cytol.*, **4**, 807–30.

RAPPORT, D. J., BERGER, J., and REID, D. B. W. (1972). Determination of food preference of *Stentor coeruleus*. *Biol. Bull.*, **142**, 103–9.

RASMUSSEN, L., and MODEWEG-HANSEN, L. (1973). Cell multiplication in *Tetrahymena* cultures after the addition of particulate material. *J. Cell Sci.*, **12**, 275–86.

RAVEN, P. H. (1970). A multiple origin for plasticids and mitochondria. *Science*, **169**, 641–6.

RENAUD, F. L., ROWE, A. J., and GIBBONS, I. R. (1968). Some properties of the protein forming the outer fibres of cilia. *J. Cell Biol.*, **36**, 79–90.

RIDDLE, J. (1962). Studies on the membrane potential of *Pelomyxa carolinensis*. *Expl. Cell. Res.*, **26**, 158–67.

ROTH, L. E. (1956). Further electron-microscope studies of *Euplotes patella*. *J. Protozool.*, **3** (Suppl.), 5.

ROUILLER, C., FAURÉ-FREMIET, E., and GAUCHERY, M. (1956a). Les tentacules d'*Ephelota;* étude au microscope electronique. *J. Protozool.*, **2**, 194–200.

ROUILLER, C., FAURÉ-FREMIET, E., and GAUCHERY, M. (1956b). Origine ciliare des fibrilles scleroproteiques pedonuclaires chez les ciliés peritriches. Étude au microscope electronique. *Expl. Cell Res.*, **11**, 527–41.

RUDZINSKA, M. A. (1956). The occurrence of hemixis in *Tokophrya infusionum*. *J. Protozool.*, **3** (suppl.), 3.

RUDZINSKA, M. A. (1958). An electron-microscope study of the contractile vacuole in *Tokophyra infusionum*. *J. biophys. biochem. Cytol.*, **4**, 195–202.

RUDZINSKA, M. A. (1961). The use of a protozoan for studies on ageing. I. Differences between young and old organisms of *Tokophyra infusionum* as revealed by light and electron microscopy. *J. Gerontology*, **16**, 213–24.

RUDZINSKA, M. A. (1965). The fine structure and function of the tentacle in *Tokophrya infusionum*. *J. Cell Biol.*, **25**, 459–77.

RUDZINSKA, M. A. (1970). The mechanism of food intake in *Tokophrya infusionum* and ultrastructural changes in food vacuoles during digestion. *J. Protozool.*, **17**, 626–41.

RUDZINSKA, M. A., JACKSON, G. J., and TUFFRAU, M. (1966). The fine

196   The Ciliates

structure of *Colpoda maupasi* with special emphasis on food vacuoles. *J. Protozool.*, **13**, 440–59.

RYLEY, J. F. (1952). Studies on the metabolism of the Protozoa. 3. Metabolism of the ciliate *Tetrahymena pyriformis* (*Glaucoma piriformis*). *Biochem. J.*, **52**, 483–92.

RYLEY, J. F. (1967). Carbohydrates and respiration. In *Chemical Zoology*, vol. 1, *Protozoa*, M. Florkin, B. T. Scheer and G. W. Kidder (Eds.), 55–92. Academic Press, London.

SANDON, H. (1932). *The Food of Protozoa*. The Egyptian University, Cairo.

SANDON, H. (1963). *Essays on Protozoology*. Hutchinson, London.

SATIR, B., SCHOOLEY, C., and SATIR, P. (1972). Membrane reorganization during secretion in *Tetrahymena*. *Nature, Lond.*, **235**, 53–4.

SATIR, P. (1965). Studies on cilia. II. Examination of the distal region of the ciliary shaft and the role of the filaments in motility. *J. Cell. Biol*, **26**, 805–34.

SCHERBAUM, O. H., and LOEFER, J. B. (1964). Environmentally induced growth oscillations in Protozoa. In *Biochemistry and Physiology of Protozoa*, vol. 3, S. H. Hutner (Ed.), 9–59. Academic Press, New York.

SCHMIDT-NEILSON, B., and SCHRAUGER, C. R. (1963). studying the contractile vacuole by micropuncture. *Science*, **139**, 606–7.

SCHNEIDER, L. (1960). Elektronmikroskopische Untersuchungen über das Nephridialsystem von *Paramecium*. *J. Protozool.*, **7**, 75–90.

SEAMAN, G. R. (1955). Metabolism in free-living ciliates. In *Biochemistry and Physiology of Protozoa*, vol. 2, S. H. Hutner and A. Lwoff (Eds.), 91–158. Academic Press, New York.

SEAMAN, G. R. (1961). Some aspects of phagotrophy in *Tetrahymena*. *J. Protozool.*, **8**, 204–12.

SEIGEL, R. W. (1963). New results on the genetics of mating types in *Paramecium bursaria*. *Genet. Res.*, **4**, 132–42.

SEIGEL, R. W. (1967). Genetics of ageing and the life cycle in ciliates. *Symp. Soc. exp. Biol.*, **21**, 127–48.

SEIGEL, R. W., and COHEN, L. W. (1963). A temporal sequence for genic expression: cell differentiation in *Paramecium*. *Am. Zool.*, **3**, 127–34.

SERAVIN, L. N. (1970). Left and right spiralling round the long body axis in ciliate Protozoa. *Acta Protozool.*, **7**, 313–23.

SHARP, R. G. (1914). *Diplodinium ecuadatum* with an account of its neuromotor apparatus. *Univ. Calif. Publ. Zool.*, **13**, 42–122.

SINDEN, R. E. (1971). The synthesis of the immobilization antigens in *Paramecium aurelia: in situ* localization of immobilization antigen using fluorescein—or ferritin—conjugated antibodies. *J. Microscopy*, **93**, 129–44.

SLEIGH, M. A. (1957). Further observations on co-ordination and the determination of frequency in the peristomal cilia of *Stentor*. *J. exp. Biol.*, **34**, 106–15.

SLEIGH, M. A. (1961). An example of metachronal co-ordination of cilia. *Nature, Lond.*, **191**, 931–2.

SLEIGH, M. A. (1962). *The Biology of Cilia and Flagella*. Pergamon Press, Oxford.

SLEIGH, M. A. (1966). The coordination and control of cilia. *Symp. Soc. exp. Biol.*, **20**, 11–31.

SLEIGH, M. A. (1968). Patterns of ciliary beating. *Symp. Soc. exp. Biol.*, **22**, 131–50.

SLEIGH, M. A. (1970). Some factors affecting the excitation of contraction in *Spirostomum*. *Acta Protozool.*, **7**, 335–52.

SLEIGH, M. A., and JARMAN, M. (1972). In preparation.

SMALL, E. B., and MARSZALEK, D. S. (1969). Scanning electron microscopy of fixed, frozen and dried Protozoa. *Science,* **163,** 1064–5.

SMITH-SONNEBORN, J., and PLAUT, W. (1967). Evidence for the presence of DNA in the pellicle of *Paramecium. J. Cell Sci.,* **2,** 225–34.

SOLDO, A. T., and VAN WAGTENDONK, W. J. (1969). The nutrition of *Paramecium aurelia,* Stock 299. *J. Protozool.,* **16,** 500–6.

SONNEBORN, T. M. (1937). Sex, sex inheritance and sex determination in *Paramecium aurelia. Proc. natn. Acad. Sci., U.S.,* **23,** 378–95.

SONNEBORN, T. M. (1939). *Paramecium aurelia:* mating types and groups; lethal interactions; determination and inheritance. *Am. Nat.,* **73,** 390–413.

SONNEBORN, T. M. (1947). Recent advances in the genetics of *Paramecium* and *Euplotes. Adv. in Genetics,* **1,** 263–358.

SONNEBORN, T. M. (1957). Breeding systems, reproductive methods and species problems in Protozoa. In *The Species Problem,* E. Mayr (Ed.), 155–324. Amer. Assoc. Adv. Sci. Publications, Washington.

SONNEBORN, T. M. (1959). Kappa and related particles in *Paramecium. Adv. Virus Res.,* **6,** 229–356.

SONNEBORN, T. M. (1963). Does preformed cell structure play an essential role in cell heredity? In *The Nature of Biological Diversity,* J. M. Allen (Ed.), 165–221. McGraw-Hill, New York.

SONNEBORN, T. M. (1970). Determination, development and inheritance of the structures of the cell cortex. In *Control Mechanisms in the Expression of Cellular Phenotypes,* H. A. Padykula (Ed.), 1–13. Academic Press, London.

STEVENS, R. E. (1970). On the apparent homology of actin and tubulin. *Science,* **168,** 845–7.

STONER, L. C., and DUNHAM, P. B. (1970). Regulations of cellular osmolarity and volume in *Tetrahymena. J. exp. Biol.,* **53,** 391–9.

SUGDEN, B., and OXFORD, A. E. (1952). Some cultural studies with holotrich ciliate Protozoa of the sheep's rumen. *J. gen. Microbiol.,* **7,** 145–53.

SUGI, H. (1960). Propagation of contraction in the stalk muscle of *Carchesium. J. Fac. Sci. Univ. Tokyo* (Sect. 4), **8,** 603–15.

SUMMERS, F. M. (1938a). Some aspects of normal development in the colonial ciliate *Zoothamnium alternans. Biol. Bull.,* **74,** 41–55.

SUMMERS, F. M. (1938b). Form regulation in *Zoothamnium alternans. Biol. Bull.,* **74,** 130–54.

TARTAR, V. (1961). *The Biology of Stentor.* Pergamon Press, Oxford.

TARTAR, V. (1967). Morphogenesis in Protozoa. In *Research in Protozoology,* vol. 2, T.-T. Chen (Ed.), 1–116. Pergamon Press, Oxford.

TAYLOR, C. V. (1920). Demonstration of the function of the neuromotor apparatus in *Euplotes* by the method of microdissection. *Univ. Calif. Publ. Zool.,* **19,** 403–71.

THOMPSON, J. C., Jr., and CORLISS, J. O. (1958). A redescription of the holotrichous ciliate *Pseudomicrothorax dubius* with particular attention to its morphogenesis. *J. Protozool.,* **5,** 175–84.

TIBBS, J. (1966). The cyst wall of *Colpoda steinii.* A substance rich in glutamic acid residues. *Biochem. J.,* **98,** 645–51.

TIBBS, J., and MARSHALL, B. J. (1970). Cyst wall protein synthesis and some changes on starvation and encystment in *Colpoda steinii. J. Protozool.,* **17,** 125–8.

TORCH, R. (1961). The nuclear apparatus of a new species of *Tracheloraphis* (Protozoa, Ciliata). *Biol. Bull.,* **121,** 410–11.

TOWNES, M. M., and BROWN, D. E. S. (1965). The involvement of pH, adenosine triphosphate, calcium and magnesium in the contraction of the glycerinated stalks of *Vorticella*. *J. cell. comp. Physiol.*, **65**, 261–9.

TUCKER, J. B. (1968). Fine structure and function of the cytopharyngeal basket in the ciliate, *Nassula*. *J. Cell Sci.*, **3**, 493–514.

TUCKER, J. B. (1971a). Spatial discrimination in the cytoplasm during microtubular morphogenesis. *Nature, Lond.*, **232**, 387–9.

TUCKER, J. B. (1971b). Development and deployment of cilia, basal bodies, and other microtubular organelles in the cortex of the ciliate *Nassula*. *J. Cell Sci.*, **9**, 539–67.

VAN WAGTENDONK, W. J. (1955). Encystment and excystment of Protozoa. In *Biochemistry and Physiology of Protozoa*, vol. 2, S. H. Hutner and A. Lwoff (Eds.), 85–90. Academic Press, New York.

VAN WAGTENDONK, W. J., CLARK, J. A. D., and GODOY, G. A. (1963). The biological status of lambda and related particles in *Paramecium aurelia*. *Proc. natn. Acad. Sci., U.S.*, **50**, 835–8.

VIVIER, É., LEGRAND, B., and PETITPREZ, A. (1969). Recherches cytochemiques et ultrastructurales sur des inclusions polysaccharidiques et calciques du spirostome; leurs relations avec la contractilité. *Protistologica*, **5**, 145–59.

WEIS-FOGH, T., and AMOS, W. B. (1972). Evidence for a new mechanism of cell motility. *Nature, Lond.*, **236**, 301–4.

WEISS, P. (1947). The problem of specificity in growth and development. *Yale J. Biol. Med.*, **19**, 235–78.

WEISZ, P. B. (1954). Morphogenesis in Protozoa. *Quart. Rev. Biol.*, **29**, 207–29.

WELLS, C. (1961). Evidence for micronuclear function during vegetative growth and reproduction of the ciliate, *Tetrahymena pyriformis*. *J. Protozool.*, **8**, 284–90.

WENYON, C. M. (1926). *Protozoology*. Baillière, Tindall & Cox, London.

WESSENBERG, H., and ANTIPA, G. (1970). Capture and ingestion of *Paramecium* by *Didinium nasutum*. *J. Protozool.*, **17**, 250–70.

WIGG, D., BOVEE, E. C., and JAHN, T. L. (1967). The evacuation mechanism of the water expulsion vesicle ('contractile vacuole') of *Amoeba proteus*. *J. Protozool.*, **14**, 104–8.

WILLIAMS, N. E. (1964). Structural development in synchronously dividing *Tetrahymena pyriformis*. In *Synchrony in Cell Division and Growth*, E. Zeuthen (Ed.), 159–75. Interscience, New York.

WILLIAMS, N. E., and SCHERBAUM, O. H. (1959). Morphogenetic events in normal and synchronously dividing *Tetrahymena*. *J. Embryol. exp. Morph.*, **7**, 241–56.

WUNDERLICH, F. (1969). The macronuclear envelope of *Tetrahymena pyriformis* GL in different physiological states. II. Frequency of central granules in pores. *Z. Zellforsch.*, **101**, 581–7.

WUNDERLICH, F., and FRANKE, W. W. (1968). Structure of macronuclear envelopes of *Tetrahymena pyriformis* in the stationary phase of growth. *J. Cell Biol.*, **38**, 458–62.

YAMAGUCHI, T. (1960). Studies on the modes of ionic behaviour across the ectoplasmic membrane of *Paramecium*. II In- and outfluxes of radioactive calcium. *J. Fac. Sci. Univ. Tokyo* (Sect. 4), **8**, 593–601.

ZAGON, I. S., VAVRA, J., and STEELE, I. (1970). Microprobe analysis of Protargol stain deposition in two Protozoa. *J. Histochem. Cytochem.*, **18**, 559–64.

INDEX

Page numbers in bold type refer to line-drawings